New Developments in the Therapy of Allergic Disorders and Asthma

International Academy for Biomedical and Drug Research
Vol. 6

Series Editors   *S.Z. Langer,* Paris
J. Mendlewicz, Brussels
G. Racagni, Milan

KARGER   Basel · Freiburg · Paris · London · New York ·
New Delhi · Bangkok · Singapore · Tokyo · Sydney

Workshop, Monte Carlo, November 9–11, 1992

New Developments in the Therapy of Allergic Disorders and Asthma

Volume Editors *S.Z. Langer,* Paris
M.K. Church, Southampton
B.B. Vargaftig, Paris
S. Nicosia, Milan

20 figures and 23 tables, 1993

Basel · Freiburg · Paris · London · New York ·
New Delhi · Bangkok · Singapore · Tokyo · Sydney

International Academy for Biomedical and Drug Research

Library of Congress Cataloging-in-Publication Data
New developments in the therapy of allergic disorders and asthma : workshop, Monte Carlo, November 9–11, 1992 / volume editors, S.Z. Langer, M.K. Church, B.B. Vargaftig, S. Nicosia.
(International Academy for Biomedical and Drug Research : vol. 6)
Includes bibliographical references and index.
1. Antiallergic agents – Congresses. 2. Antiasthmatic agents – Congresses. I. Langer, S.Z. (Salomon Z.)
II. Church, M.K. (Martin K.)   III. Vargaftig, B.B. (Boris B.)   IV. Nicosia, S. (Simonetta)
V. Series: International Academy for Biomedical and Drug Research (Series) ; vol. 6.
[DNLM: 1. Asthma – drug therapy – congresses. 2. Hypersensitivity – drug therapy – congresses.
W1 IN701D v. 6 1993 / WD 300 N5315 1992]
ISBN 3-8055-5748-5 (alk. paper)

Drug Dosage. The authors and the publisher have exerted every effort to ensure that drug selection and dosage set forth in this text are in accord with current recommendations and practice at the time of publication. However, in view of ongoing research, changes in government regulations, and the constant flow of information relating to drug therapy and drug reactions, the reader is urged to check the package insert for each drug for any change in indications and dosage and for added warnings and precautions. This is particularly important when the recommended agent is a new and/or infrequently employed drug.

All rights reserved. No part of this publication may be translated into other languages, reproduced or utilized in any form or by any means, electronic or mechanical, including photocopying, recording, microcopying, or by any information storage and retrieval system, without permission in writing from the publisher.

© Copyright 1993 by S. Karger AG, P.O. Box, CH–4009 Basel (Switzerland)
Printed in Switzerland on acid-free paper by Thür AG Offsetdruck, Pratteln
ISBN 3-8055-5748-5

Contents

Molecular Pharmacology of Histamine Receptors 1
    Traiffort, E.; Arrang, J.-M.; Garbarg, M.; Ruat, M.; Leurs, R.;
    Tardivel-Lacombe, J.; Rouleau, A.; Schwartz, J.-Ch. (Paris)

Immunological Aspects of the Pathogenesis of Asthma 10
    Corrigan, C.J.; Kay, A.B. (London)

Evidence for a Biological Activity of Anti-Tissue Antisera on an Isolated Cell System   20
    Bobo, M.-H.; Magous, R. (Montpellier); Pouderoux, P.; Balmes, J.-L. (Nîmes);
    Mingard, P. (Lausanne); Bali, J.-P. (Montpellier)

Mechanisms of Experimental Bronchopulmonary Hyperresponsiveness as Related to
    Eosinophils 27
    Vargaftig, B.B. (Paris)

Acetaldehyde Induces a Bronchoconstrictor Response in Guinea Pigs. A Pharmacological Study 33
    Berti, F.; Rossoni, G.; Buschi, A.; Robuschi, M.; Trento, F.; Della Bella, D.
    (Milan)

Physical and Chronic Idiopathic Urticaria 44
    Juhlin, L. (Uppsala)

H_1-Antihistamines as Broad-Spectrum Drugs for the Treatment of Various Allergic
    Disorders 55
    Rihoux, J.-P. (Braine-l'Alleud)

Specific Immunotherapy 60
    Bousquet, J.; Michel, F.-B. (Montpellier)

Inhibitors of Leukotriene Synthesis and Actions: Asthma Drugs of the Near Future? . 71
    McMillan, R.M. (Macclesfield, Cheshire)

Binding Sites for Peptido-Leukotrienes in Human Lung Parenchyma 86
    Nicosia, S.; Capra, V.; Giovanazzi, S.; Rovati, G.E. (Milan)

Effects of Antidiencephalon Antibody (Ser 282) on Sleep in Primary Fibromyalgia.
A Preliminary Report 91
    Staner, L.; Kempenaers, Ch.; Appelboom, T. (Brussels); Mingard, P. (Lausanne);
    Mendlewicz, J. (Brussels)

Basic and Clinical Aspects of Atopic Dermatitis 96
    Rajka, G. (Oslo)

The Actions of Antihistamines in Allergic Disease 107
    Church, M.K. (Southampton)

New Targets for Antiallergic Agents 115
    Ciprandi, G.; Buscaglia, S.; Pronzato, C.; Ricca, V.; Pesce, G.P.; Villaggio, B.;
    Fiorino, N.; Albano, M.; Scordamaglia, A.; Bagnasco, M.; Canonica, G.W.
    (Genoa)

Current Issues in the Therapy of Allergy and Asthma. Panel Discussion 128

    Subject Index 133

Molecular Pharmacology of Histamine Receptors

Elisabeth Traiffort, Jean-Michel Arrang, Monique Garbarg, Martial Ruat, Rob Leurs, Joël Tardivel-Lacombe, Agnès Rouleau, Jean-Charles Schwartz

Unité de Neurobiologie et Pharmacologie (U–109) de l'INSERM, Centre Paul-Broca, Paris, France

The early history of histamine is largely associated with allergy. The major actions of histamine were described at the beginning of this century by Sir Henry Dale and his colleagues after its isolation from ergot extracts. Namely its potent contractile effects on smooth muscles and the capillary dilation it induces, which mimic some initial manifestations of the anaphylactic shock, were identified by these scientists. They also detected the presence of the amine in many tissues but it is another British scientist, Feldberg, who clearly demonstrated that histamine was released from the lung during the anaphylactic response and that it induced a marked bronchoconstriction.

The idea that histamine exerts its various biological effects via interaction with several distinct receptor subtypes progressively arose mainly with the design of subtype-selective antagonists. It was first realized that the 'antihistamines' (now termed H_1-receptor antagonists), the first of which were designed by Bovet and Staub [1], did not block uniformly all actions of histamine, leaving, for instance, gastric secretion unaffected. On this basis as well as on the differential action of agonists, Ash and Schild [2] clearly postulated the existence of a second receptor subtype. The existence of the H_2 receptor was definitively established with the design of burimamide, a selective (non-H_1) antagonist, as well as of several relatively selective agonists [3].

Arrang et al. [4] proposed the existence of the third receptor subtype, an autoreceptor controlling the synthesis and release of histamine in cerebral neurons. Four years later, the existence of the H_3 receptor was definitively established with the design of highly potent and selective agonists and antagonists [5].

The field of histamine receptor pharmacology and biochemistry was recently reviewed in an extensive manner [6–8]. However, the very recent cloning of the genes encoding two of the histamine receptor subtypes has notably enlarged our knowledge of these receptors.

The Histamine H_1 Receptor

Biochemical and localisation studies of the H_1 receptor were made feasible with the design of reversible and irreversible radiolabelled probes such as [^{3}H]mepyramine, [^{125}I]iodobolpyramine and [^{125}I]iodoazidophenpyramine [9–11].

Initial biochemical studies indicated that the cerebral guinea pig H_1 receptor was a glycoprotein of apparent molecular size of 56 kD with critical disulfide bonds and that agonist binding was regulated by guanyl nucleotides, implying that the receptor belongs to the superfamily of receptors coupled to G proteins. In addition, various intracellular responses were found to be associated with H_1-receptor stimulation, e.g. inositol phosphate release, increase in Ca^{2+} fluxes, cyclic AMP and cyclic GMP accumulation in whole cells, arachidonic acid release [6, 12]. It was not known, however, whether such a variety of responses corresponds to a single receptor or to distinct isoreceptor. Indeed several photoaffinity-labelled proteins of slightly different sizes, but similar H_1 pharmacology, were detected in some tissues [10].

In spite of preliminary attempts using affinity columns with a mepyramine derivative, the H_1 receptor was never purified to homogeneity. Nevertheless, the deduced amino acid sequence of a bovine H_1 receptor was recently disclosed after expression cloning of a corresponding cDNA. The latter was based upon the detection of a Ca^{2+}-dependent Cl^- influx into microinjected *Xenopus* oocytes. Following the transient expression of the cloned cDNA into COS-7 cells, the identity of the protein as an H_1 receptor was confirmed by binding studies [13].

In order to identify the signalling systems of the H_1 receptor in a well-studied animal species as well as to assess the possible existence of isoreceptors, we have recently cloned a guinea pig cDNA encoding a H_1 receptor starting from the bovine sequence [14]. It encodes a glycoprotein of 488 amino acids (fig. 1) with a calculated M_r of 56 kD, in good agreement with the apparent size of the photoaffinity-labelled receptor from guinea pig brain or heart, as determined by SDS/PAGE analysis [15, 16]. Southern blot analysis, using a variety of restriction enzymes, did not provide any evidence for multiple H_1 isoreceptors and Northern blot analysis of a variety of guinea pig peripheral or cerebral tissues identified, in most cases, a single transcript of 3.3 kb. There was, however, in some tissues, e.g. ileum or lung, several transcripts possibly generated by the use of distinct promoter or polyadenylation sites. Nevertheless, one cannot exclude the existence of dis-

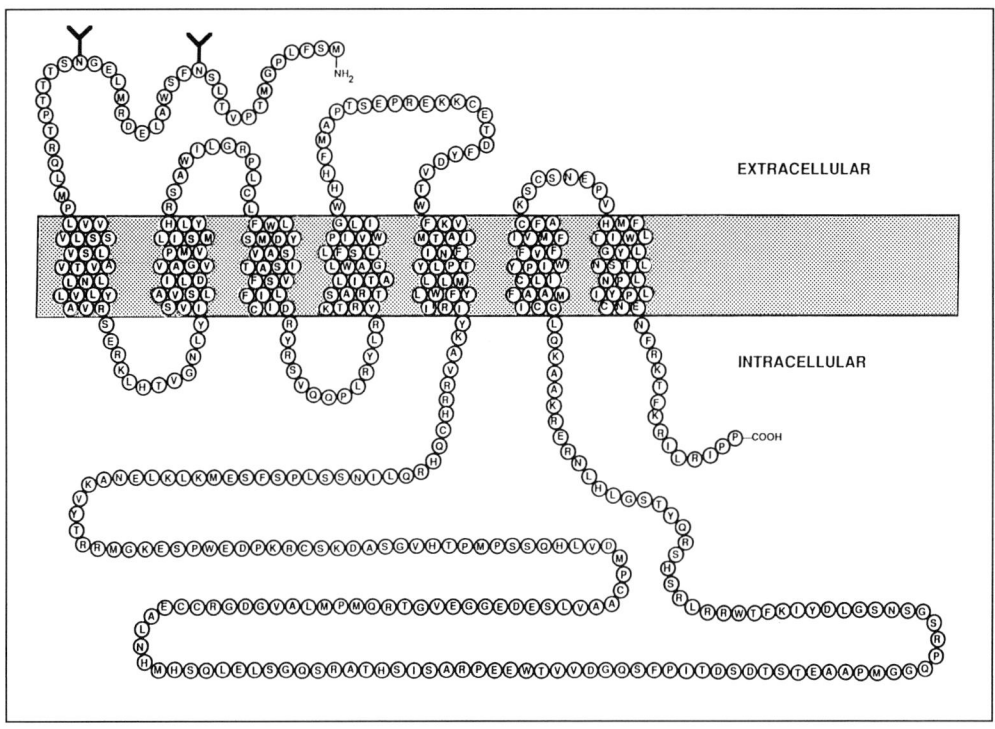

Fig. 1. Amino acid sequence of the guinea pig H_1 receptor. Y indicates the presence of glycosylation sites.

tinct genes, although the sizes of transcripts in brain and heart were the same, contrasting with the existence of distinct photoaffinity labelled proteins in these tissues [16].

In situ hybridization studies showed a highly contrasted expression of the H_1 receptor gene transcript in guinea pig brain [14]. When compared with the autoradiographic localisation of the corresponding receptor protein [17], consistent as well as complementary information was provided. For instance the mRNAs were found in high levels in cerebellar Purkinje cells and hippocampal pyramidal cells, whereas dense [^{125}I]iodobolpyramine binding sites are found in the molecular layers of both areas: this presumably reflects the synthesis of the receptor in perikarya and its final insertion in membranes of the abundant dendritic trees of both cell types.

Transfection with the guinea pig gene followed by stable expression of H_1 receptors by a CHO cell line allowed the characterization of multiple signalling

Table 1. Summary of the various responses mediated by the H_1 receptor in CHO(H_1) and HeLa cells

Response	Cell line	EC$_{50}$ of histamine µM	Sensitivity to [Ca]$_e$	PTX	PMA
Inhibition of [^{3}H]mepyramine binding	CHO(H_1)	40±6[1]	–	–	–
Inositol phosphate mobilization	CHO(H_1)	1.4±0.1	partial	no	partial
Intracellular Ca^{2+} increase	CHO(H_1)	~1	no[2]	no	no
cAMP increase (forskolin potentiation)	CHO(H_1)	0.4±0.1	no	no	±
Arachidonate increase	CHO(H_1)	1.3±0.2	yes	partial	yes
Arachidonate increase	HeLa	1.9±0.2	yes	no	yes

[1] Evaluated in the absence of added guanyl nucleotide.
[2] Initial peak insensitive but sustained late phase is [Ca^{2+}]$_e$-dependent.

pathways (table 1) [18]. In each case we assessed the involvement of a G_i/G_o protein with pertussis toxin (PTX), the influence of extracellular Ca^{2+} and of protein kinase C (PKC) activation by phorbol-12-myristate-13-acetate (PMA).

Histamine induced in PTX– and PMA-insensitive manner a biphasic increase in intracellular Ca^{2+} level of which only the second, sustained phase was dependent on the extracellular Ca^{2+} level.

In addition, histamine also caused a 3-fold elevation of inositol phosphate production which was PTX-insensitive but slightly inhibited by PMA and reduced by 75% in the absence of extracellular Ca^{2+}.

Histamine also caused a massive release of arachidonic acid (AA), occurring in a Ca^{2+}- and PMA-sensitive manner probably through the activation of a cytosolic phospholipase A$_2$, which partly involves coupling to a PTX-sensitive G protein. In comparison, in HeLa cells endowed with a native H_1 receptor, the histamine-induced arachidonic acid release was also Ca^{2+}- and PMA-sensitive but totally PTX-insensitive.

Finally, in the same CHO(H_1) cell line, histamine in very low concentrations potentiated the cyclic AMP accumulation induced by forskolin. This response appeared to be insensitive to PTX, extracellular calcium and PMA.

These various observations show that stimulation of a single receptor subtype, the guinea pig H_1 receptor, can trigger four major intracellular signals, presumably through coupling to several G proteins which are variously modulated by extracellular Ca^{2+} and PKC activation.

The Histamine H_2 Receptor

Molecular properties of the H_2 receptor have remained largely unknown for a long time. For instance it is only recently that reversible labelling of the H_2 receptor was achieved using [^{3}H]tiotidine [19] or, more reliably, [^{125}I]iodoaminopotentidine [20]. Irreversible labelling, achieved with a photoaffinity probe, followed by SDS-PAGE led to the identification of H_2 receptor peptides from the guinea pig brain [20].

By screening cDNA or genomic libraries with homologous probes, the gene encoding the H_2 receptor was first identified in dog [21] and, then, in rat [22] and man [23]. Comparison of these proteins shows that they display a high degree of homology, i.e. 82% between the rat and dog receptor, whereas the degree of homology between the H_1 and H_2 receptors is limited. The H_2 receptor is organized like other receptors positively coupled to adenylyl cyclase, i.e. it displays a short (30 amino acids) third intracellular loop and a long (71 amino acids) C-terminal cytoplasmic tail.

Consistent with their histamine binding function, the H_2 receptors display in the third transmembrane helix (TM3) an aspartate residue (Asp^{98}) likely to bind the ammonium group of the endogenous ligand, as it is found in all other aminergic receptors. In the TM5 an aspartate and a threonine residue (Asp^{185} and Thr^{189} in the rat and Asp^{186} and Thr^{190} in the dog) seemed responsible for hydrogen bonding with the nitrogen atoms of the imidazole ring of histamine. This was recently confirmed by site-directed mutagenesis [24].

A potential regulation of the rat H_2 receptor by phosphorylation is suggested by the presence of three consensus sites for protein kinase C.

Northern blot analysis of various tissues using a probe derived from the rat cDNA sequence revealed the presence of a single major transcript of 6.0 and 4.5 kb in rat and guinea pig, respectively [22, 25]. The distribution of the mRNAs in these two species was consistent with the known distribution of the receptor as mainly established using the sensitive probe [^{125}I]iodoaminopotentidine [20].

Transfected CHO cells were found to express a high level of rat H_2 receptors [25]. In these cells, histamine, in low concentration, induced an accumulation of cAMP, confirming the association of the H_2 receptor with adenylyl cyclase. In addition, in the same cells, histamine potently inhibited the release of arachidonic acid induced by stimulation of constitutive purinergic receptors or by application

Table 2. Potent and selective H_3 receptor agonists

Compound	Relative potency at receptors		
	H_1	H_2	H_3
Histamine: Im–CH_2–CH_2–NH_2	100	100	100
(R)α-methylhistamine: Im–CH_2–CH(CH_3)–NH_2	0.5	1	1,500
(R)α,(S)β-dimethylhistamine: Im–CH(CH_3)–CH(CH_3)–NH_2	0.03	0.2	1,800
Imetit: Im–CH_2–CH_2–S–C(=NH)–NH_2	<0.1	0.6	6,200

of a Ca^{2+} ionophore. This inhibition was independent of either cAMP or Ca^{2+} levels in the cells. The results indicated that a single H_2 receptor may be linked not only to adenylyl cyclase activation but also to reduction of phospholipase A_2 activity. Because H_1 receptors have been reported to stimulate arachidonic acid release, inhibition of this release, an unexpected signalling pathway for H_2 receptors, may account for the opposite physiological responses elicited in many tissues by activation of these two receptor subtypes.

The Histamine H_3 Receptor

The H_3 receptor was initially detected as an autoreceptor controlling histamine synthesis and release in brain [4]. It was thereafter shown to inhibit presynaptically the release of other monoamines in brain and peripheral tissues as well as of neuropeptides from unmyelinated C fibers [7, 26].

The molecular structure of the H_3 receptor remains to be established. Reversible labelling of this receptor was first achieved using a highly selective agonist [^{3}H](R)α-methylhistamine [5], then [^{3}H]Nα-methylhistamine, a less selective agonist was also proposed [27] as well as, more recently, [^{125}I]iodophenpropit [28]. It appears that the binding of [^{3}H](R)α-methylhistamine is regulated by guanyl nucleotides, strongly suggesting that the H_3 receptor, like the other histamine receptors, belongs to the superfamily of receptors coupled to G proteins [29]. Con-

Table 3. Effects of (R)α–methylhistamine, an H_3 receptor agonist, on [^{3}H]histamine synthesis in lung of capsaicin-pretreated rats

	[^{3}H]Histamine synthesis, dpm/mg	
	controls	capsaicin
Vehicle	5.6 ± 0.5	5.1 ± 0.4
(R)α-methylhistamine	3.9 ± 0.3* (−30%)	4.8 ± 0.3 (n.s.)

Four weeks after neonatal capsaicin groups of 13–44 rats received (R)α-methylhistamine (15 mg/kg i.p.) and were killed 3 h later. They had received 250 μCi of [^{3}H]*L*-histidine i.v. 10 min before sacrifice. *p < 0.05.

stitutive H_3 receptors in a gastric cell line appear to be negatively coupled to phospholipase C [30].

During the last years several potent and highly selective H_3 receptor agonists were designed [5, 9, 31], two of which bear asymmetric centers and display a high degree of stereoselectivity (table 2).

Using these compounds as well as the H_3 receptor antagonist thioperamide [5], several novel actions and physiological roles of histamine could be unraveled.

In the field of airway inflammation, the work of Prof. Barnes and colleagues has clearly established the multiple inhibitory effect of H_3 receptor stimulation in airways [32–35]. H_3 receptors seem to be located presynaptically on preganglionary and postganglionary cholinergic fibers and, thereby, to modulate in an inhibitory fashion acetylcholine release. In addition, and quite interestingly, H_3 receptors inhibit presynaptically the release of pro-inflammatory tachykinins from unmyelinated C fibers in airways.

It seems likely that such an effect is induced through release of histamine from mast cells which are abundant in airways and in close contact with peptidergic C fibers. It is likely, in turn, that these fibers control mast cell functions since the H_3 receptor-mediated modulation of [^{3}H]histamine synthesis in rat lung is no longer apparent in rats receiving capsaicin neonatally to selectively destroy these fibers (table 3).

It is therefore apparent that mast cells and peptidergic C fibers control each other through a short feedback loop involving H_3 receptors. In view of the large repertoire of mast cells in inflammatory mediators, it can be reasoned that inflammatory airway diseases such as asthma may involve a dysregulation of this loop. The hypothesis that H_3 receptor agonists may represent a new class of anti-inflammatory agents remains to be assessed.

References

1 Bovet D, Staub AM: Action protectrice des éthers phénoliques au cours de l'intoxication histaminique. CR Soc Biol Paris 1937;124:547–549.
2 Ash ASF, Schild HO: Receptors mediating some actions of histamine. Br J Pharmacol Chemother 1966;27:427–439.
3 Black JW, Duncan WAM, Durant CJ, Ganellin CR, Parsons ME: Definition and antagonism of histamine H_2-receptors. Nature 1972;236:385–390.
4 Arrang JM, Garbarg M, Schwartz JC: Autoinhibition of histamine release mediated by a novel class (H_3) of histamine receptor. Nature 1983;302:832–837.
5 Arrang JM, Garbarg M, Lancelot JC, Lecomte JM, Pollard H, Robba M, Schunack W, Schwartz JC: Highly potent and selective ligands for histamine H_3-receptors. Nature 1987;327:117–123.
6 Hill SJ: Distribution, properties and functional characteristics of three classes of histamine receptors. Pharmacol Rev 1990;42:46–83.
7 Schwartz JC, Arrang JM, Garbarg M, Pollard H, Ruat M: Histaminergic transmission in the mammalian brain. Physiol Rev 1991;71:1–51.
8 Schwartz JC, Haas HL (eds): The Histamine Receptor. Receptor Biochemistry and Methodology. New York, Wiley/Liss, 1992, vol 16.
9 Garbarg M, Traiffort E, Ruat M, Arrang JM, Schwartz JC: Reversible labelling of H_1, H_2 and H_3-receptors; in Schwartz JC, Haas HL (eds): The Histamine Receptor. New York, Wiley/Liss, 1992, pp 73–95.
10 Ruat M, Traiffort E, Schwartz JC: Biochemical properties of histamine receptors; in Schwartz JC, Haas HL (eds): The Histamine Receptor. New York, Wiley/Liss, 1992, pp 97–107.
11 Pollard H, Bouthenet ML: Autoradiographic visualization of the three histamine receptor subtypes in the brain; in Schwartz JC, Haas HL (eds): The Histamine Receptor. New York, Wiley/Liss, 1992, pp 179–192.
12 Hill SJ, Donaldson J: The H_1 receptor and inositol phospholipid hydrolysis; in Schwartz JC, Haas HL (eds): The Histamine Receptor. New York, Wiley/Lyss, 1992, pp 109–128.
13 Yamashita M, Fukui H, Sugawa K, Horio Y, Ito S, Mizuguchi H, Wada H: Expression cloning of a cDNA encoding the bovine histamine H_1 receptor. Proc Natl Acad Sci USA 1991;88:11515–11519.
14 Traiffort E, Leurs R, Arrang JM, Tardivel-Lacombe J, Diaz J, Schwartz JC, Ruat M: Guinea pig histamine H_1 receptor. I. Gene cloning, characterization and tissue expression revealed by in situ hybridization. J Neurochem, in press.
15 Ruat M, Körner M, Garbarg M, Gros C, Schwartz JC, Tertiuk W, Ganellin CR: Characterization of histamine H_1-receptor binding peptides in guinea pig brain using [^{125}I]iodoazidophenpyramine, an irreversible specific photoaffinity probe. Proc Natl Acad Sci USA 1988;85:2743–2747.
16 Ruat M, Bouthenet ML, Schwartz JC, Ganellin CR: Histamine H_1 receptor in heart: Unique electrophoretic mobility and autoradiographic localization. J Neurochem 1990;55:379–385.
17 Bouthenet ML, Ruat M, Salès N, Garbarg M, Schwartz JC: A detailed mapping of histamine H_1-receptors in guinea pig central nervous system established by autoradiography with [^{125}I]iodobolpyramine. Neuroscience 1988;26:553–600.

18 Leurs R, Traiffort E, Arrang JM, Tardivel-Lacombe J, Ruat M, Schwartz JC: Guinea pig histamine H_1 receptor. II. Stable expression in Chinese hamster ovary cells reveals the interaction with three major signal transduction pathways. J Neurochem, in press.

19 Gajtkowski GA, Norris DB, Rising TJ, Wood TP: Specific binding of ^{3}H-tiotidine to histamine H_2-receptors in guinea pig cerebral cortex. Nature 1983;304:65–67.

20 Ruat M, Traiffort E, Bouthenet ML, Schwartz JC, Hirschfeld J, Buschauer A, Schunack W: Reversible and irreversible labeling and autoradiographic localization of the cerebral histamine H_2 receptor and [^{125}I]iodinated probes. Proc Natl Acad Sci USA 1990;87:1658–1662.

21 Gantz I, Schaffer M, Delvalle J, Logsdon C, Campbell V, Uhler M, Yamada T: Molecular cloning of a gene encoding the histamine H_2 receptor. Proc Natl Acad Sci USA 1991;88:429–433.

22 Ruat M, Traiffort E, Arrang JM, Leurs R, Schwartz JC: Cloning and tissue expression of a rat histamine H_2-receptor gene. Biochem Biophys Res Commun 1991;179:1470–1478.

23 Gantz I, Munzert G, Tashiro T, Schaffer M, Wang L, Delvalle J, Yamada T: Molecular cloning of the human histamine H_2 receptor. Biochem Biophys Res Commun 1991;178:1386–1392.

24 Gantz I, Delvalle J, Wang LD, Tashiro T, Munzert G, Guo YJ, Konda Y, Yamada T: Molecular basis for the interaction of histamine with the histamine H_2 receptor. J Biol Chem 1992;267:20840–20843.

25 Traiffort E, Ruat M, Arrang JM, Leurs R, Piomelli D, Schwartz JC: Expression of a cloned rat histamine H_2 receptor mediating inhibition of arachidonate release and activation of cAMP accumulation. Proc Natl Acad Sci USA 1992;89:2649–2653.

26 Arrang JM, Garbarg M, Schwartz JC: H_3-receptor and control of histamine release; in Schwartz JC, Haas HL (eds): The Histamine Receptor. New York, Wiley/Liss, 1992, pp 145–159.

27 Korte A, Myers J, Shih NY, Egan RW, Clark MA: Characterization and tissue distribution of H_3-histamine receptors in guinea pigs by N^α-methylhistamine. Biochem Biophys Res Commun 1990; 168:979–986.

28 Jansen FP, Rademaker B, Bast A, Timmerman H: The first radiolabeled histamine H_3 receptor antagonist, [^{125}I]iodophenpropit: Saturable and reversible binding to rat cortex membranes. Eur J Pharmacol 1992;217:203–205.

29 Arrang JM, Roy J, Morgat JL, Schunack W, Schwartz JC: Histamine H_3-receptor binding sites in rat brain membranes: Modulation by guanine nucleotides and divalent cations. Eur J Pharmacol Mol Pharmacol Sect 1990;188:219–227.

30 Cherifi Y, Pigeon C, Le Romancer M, Bado A, Reyl-Desmars F, Lewin MJM: Purification of a histamine H_3 receptor negatively coupled to phosphoinositide turnover in the human gastric cell line HGT1. J Biol Chem 1992;267:25315–25320.

31 Lipp R, Stark H, Schunack W: Pharmacochemistry of H_3 receptors; in Schwartz JC, Haas HL (eds): The Histamine Receptor. New York, Wiley/Liss, 1992, pp 57–72.

32 Ichinose M, Stretton CD, Schwartz JC, Barnes PJ: Histamine H_3-receptors inhibit cholinergic neurotransmission in guinea pig airways. Br J Pharmacol 1989;97:13–15.

33 Ichinose M, Barnes PJ: Inhibitory histamine H_3-receptors on cholinergic nerves in human airways. Eur J Pharmacol 1989;163:383–386.

34 Ichinose M, Barnes PJ: Histamine H_3-receptors modulate non-adrenergic, non-cholinergic bronchoconstriction in guinea pig in vivo. Eur J Pharmacol 1989;174:49–55.

35 Barnes PJ: Histamine receptors in the respiratory tract; in Schwartz JC, Haas HL (eds): The Histamine Receptor. New York, Wiley/Liss, 1992, pp 253–270.

Elisabeth Traiffort, Unité de Neurobiologie et Pharmacologie (U–109) de l'INSERM,
Centre Paul-Broca, 2ter, rue d'Alésia, F–75014 Paris (France)

Immunological Aspects of the Pathogenesis of Asthma

C.J. Corrigan, A.B. Kay

Department of Allergy and Clinical Immunology, National Heart and Lung Institute, London, UK

Introduction

It is now widely accepted that chronic mucosal inflammation plays an important role in the pathogenesis of asthma, despite the fact that the precise relationship of inflammation to symptoms, or to objective measures of disease severity, remains unclear. It is the aim of this article briefly to describe the immunobiology of T lymphocytes and the evidence that they play a role in asthma. It should be borne in mind that inflammatory cells do not act in isolation but interact both with each other and with resident tissue cells and local neural networks.

Evidence for Mucosal Inflammation in Asthma

The classical studies of bronchial histopathology in patients having died of asthma [1–3] showed an intense invasion of the bronchial mucosa with inflammatory cells, particularly eosinophils, macrophages and lymphocytes. Neutrophils were present but in fewer numbers. Deposition of eosinophil products in and around the bronchial epithelium was a particularly prominent feature [4]. Other features noted in these studies which appear to be typical of asthma include loss of surface lining epithelium, collagen deposition in the reticular layer beneath the basement membrane of the epithelium [5, 6], dilatation of blood vessels, mucosal oedema and hypertrophy of both submucosal glands and bronchial smooth muscle.

More recent studies have been performed on mild asthmatic volunteers, utilising the techniques of bronchoalveolar lavage (BAL) and bronchial biopsy through the flexible fibreoptic bronchoscope. These studies have shown that many of the inflammatory changes observed in asthma deaths are also a feature of milder, apparently well-controlled disease. Elevated numbers of eosinophils, both in the bronchial mucosa and in BAL fluid, were constant features of mild asthma [7–9]. Similarly, increased numbers of activated lymphocytes, identified either as irregular, atypical lymphocytes by transmission electron microscopy [10] or as CD25 (IL-2 receptor)-bearing cells as shown by immunocytochemistry [7] were also invariably seen. There exists evidence that activation of T lymphocytes and subsequent eosinophil recruitment and secretion may contribute to epithelial damage and possibly also to bronchial hyperresponsiveness in asthma [11–14]. In contrast, recent studies of the bronchial mucosa in patients with mild asthma associated with atopy demonstrated no significant changes in the numbers of mast cells or of their subtypes in the bronchial mucosa [7, 15]. Similarly, bronchial mucosal neutrophils were not increased in number in these patients.

Asthma has been traditionally subdivided clinically according to its apparent aetiology (intrinsic, extrinsic and occupational), implying possible variability in its pathogenesis. One important question therefore relates to whether these clinical distinctions are apparent in histopathological terms. Preliminary studies addressing this question would suggest that they are not: an autopsy study of the bronchial mucosa of a patient who had died with severe occupational asthma showed histological changes similar to those seen in fatal non-occupational asthma [16], while a recent immunocytochemical study [17] comparing bronchial biopsies from extrinsic and occupational asthmatics showed that these were indistinguishable in terms of their inflammatory cell infiltrate: in both cases the biopsies showed increased numbers of activated eosinophils and T lymphocytes, but not neutrophils and monocyte/macrophages, as compared to biopsies from normal control subjects. A similar situation pertained with 'intrinsic' asthmatics, although there was in this case some evidence of an additional macrophage infiltrate [18]. Examination of BAL fluid obtained from a group of 'intrinsic' asthmatics [19] showed increased numbers of activated T lymphocytes, eosinophils and neutrophils as compared to normal controls. These observations suggest that the bronchial response in patients with asthma is uniform, regardless of the nature of the provoking agent, and lend support to the hypothesis that the pathogenesis of asthma is independent of coexisting atopy.

Immunobiology of T Lymphocytes and Evidence for Their Involvement in Asthma

Pro-Inflammatory Role of CD4 T Lymphocytes
CD4 T lymphocytes have a central role to play in any antigen-driven inflammatory process (cell or antibody mediated), since they are the only cells capable or recognizing 'foreign' antigenic material after processing by antigen-presenting

cells. It is now clear that CD4 T lymphocytes, after activation by antigen, have the capacity to elaborate a wide variety of protein mediators called lymphokines. These lymphokines have the capacity to orchestrate the differentiation, recruitment, accumulation and activation of specific granulocyte effector cells at mucosal surfaces. In the case of eosinophils, CD4 T lymphocytes are a major source of IL-5 which was demonstrated to: (a) promote the differentiation of mature eosinophils from precursor cells [20, 21]; (b) prolong the survival of eosinophils in vitro from days to weeks, especially in the presence of fibroblasts or endothelial cells [22, 23]; (c) exhibit chemotactic activity for eosinophils but not neutrophils in vivo, although this effect was weak and requires further confirmation [24]; (d) enhance the adhesion of eosinophils, but not neutrophils to vascular endothelial cells, a vital initial step in tissue emigration [25]; (e) prime eosinophils for increased activity in a number of subsequent effector responses, including antibody-mediated killing of parasitic larvae, elaboration of lipid mediators and activation by PAF [23, 26].

Similar effects on eosinophils were exhibited by IL-3 [27] and GM-CSF [28, 29], although unlike IL-5 these lymphokines are not eosinophil specific. Interferon-γ was shown to enhance eosinophil cytotoxicity [30]. The observations that T lymphocyte clones from patients with the hypereosinophilic syndrome demonstrated IL-5-like activity [31] and that cultured T lymphocytes from patients with asthma spontaneously secreted lymphokines which could prolong eosinophil survival [32] directly support the hypothesis that eosinophil numbers and function may be regulated by T lymphocytes in vivo.

These experiments emphasize the facts that activated CD4 T lymphocytes have the propensity to bring about *selective* accumulation and activation of specific granulocytes in tissues, and that T lymphocyte-mediated granulocyte accumulation and activation need not be dependent on the presence of antibodies, including IgE.

It is worth bearing in mind that T lymphocytes are the principal but not the only source of lymphokines. Eosinophils and mast cells have also been shown to produce a variety of lymphokines [33, 34].

Th1 and Th2 CD4 T Lymphocytes

At the present time not all lymphokines have been implicated in the pathogenesis of asthmatic inflammation. Il-3, IL-5 and GM-CSF are strongly implicated in that they can selectively recruit and activate mast cells and eosinophils, and IL-4 is implicated in the sense that it is responsible for the promotion of inappropriate IgE synthesis and the consequences thereof. The genes encoding these lymphokines are located relatively close together in the human genome, on the long arm of chromosome 5, raising the possibility that their expression may be at least in part coordinately regulated. There is already good evidence that this is the case in mouse T lymphocytes. Antigen-activated murine CD4 T lymphocyte clones can be divided into two broad types, called Th1 and Th2, according to the pattern of lymphokines they secrete [35]. Th1 cells secrete IL-2, interferon-γ and TNF-β, but not IL-4, IL-5 and IL-6. Th2 cells secrete IL-4, IL-5 and IL-6 but not IL-2, interferon-γ and TNF-β. Other lymphokines, including IL-3 and GM-CSF are secreted by both cell types.

The functional capacities of Th1 and Th2 CD4 T lymphocyte clones differ in a manner which reflects their respective patterns of lymphokine synthesis. Th2 clones, through their secretion of IL-4 and IL-5, serve as excellent helper cells for the synthesis of immunoglobulins including IgE by B lymphocytes in vitro [36], since both these lymphokines non-specifically enhance B lymphocyte activation. In addition, by their secretion of IL-3, IL-4 and IL-5, Th2 clones could activate mast cells and eosinophils, and are therefore strongly implicated in the pathogenesis of allergic and asthmatic inflammation. Th1 clones have been shown to provide help for B lymphocyte IgG synthesis [37], but they strongly suppress IgE production through their release of interferon-γ which also suppresses B lymphocyte proliferation in a non-specific fashion [38]. Th1 cells, but not Th2 cells were also shown to have the capacity to elicit delayed-type hypersensitivity (DTH) reactions in vivo [39].

In contrast to murine cells, human CD4 T lymphocyte clones stimulated at random using lectins do not fall cleanly into Th1 and Th2 patterns, and there are many examples of clones which secrete a mixture of lymphokines characteristic of both categories [40]. Nevertheless, T lymphocytes with Th1 and Th2 type patterns of lymphokine secretion do appear to exist in vivo (see below). These data can be reconciled with those from mice if it is assumed that precursors of Th1 and Th2 cells (Th0 cells) exist which secrete a mixture of Th1 and Th2 type lymphokines, and that these Th0 cells develop into Th1 or Th2 cells under the influence of extraneous factors such as their antigen specificity, the site of antigen presentation and the nature of the antigen-presenting cells.

Experimental Observations Implicating Activated CD4 T Lymphocytes in the Pathogenesis of Asthma

In several recent studies of bronchial biopsies obtained from asthmatics of varying clinical aetiology [7, 15, 17, 18], the numbers and activation status of mucosal T lymphocytes were assessed by immunostaining with monoclonal antibodies directed against T lymphocyte phenotypic and activation markers. Interestingly, the total numbers of both CD4 and CD8 T lymphocytes in the bronchial mucosa of these asthmatics were not significantly elevated as compared to normal controls; CD4 cells predominated over CD8 in both cases. In contrast, only cells in the biopsies from asthmatics showed evidence of IL-2 receptor expression, suggesting activation. Furthermore, in the biopsies from asthmatics, the numbers of activated T lymphocytes could be correlated with both the total numbers of eosinophils and the numbers of activated eosinophils. Finally, the degree of activation could be correlated with disease severity, as assessed by measurement of bronchial hyperresponsiveness. These observations provide circumstantial evidence supporting the hypothesis that activated CD4 T lymphocytes control the numbers and activation status of eosinophils in asthmatic bronchial inflammation, and that the degree of activation is one factor which determines disease severity. Using immunostaining and flow cytometry, it was shown that a proportion of CD4 T lymphocytes, but not CD8 cells, in the peripheral blood of patients with acute severe asthma are activated, as assessed by expression of IL-2 receptor, HLA-DR and VLA-1 [41]. The degree of activation of these cells decreased after therapy to an extent that could be correlated with the degree of clinical improvement [42].

Some studies demonstrated an increase in the relative numbers of lymphocytes found in BAL fluid obtained from patients with mild, stable asthma [43], whereas others showed similar numbers in asthmatics and normal controls [8]. A recent flow cytometric study on T lymphocytes in BAL fluid of mild atopic asthmatics [44] showed evidence of increased expression of activation markers on both CD4 and CD8 cells, although only activated CD4 cell numbers correlated with the numbers of BAL eosinophils. In a further study employing allergen bronchial challenge of sensitized atopic asthmatics [45], a selective increase in CD4 cells in BAL fluid was observed 48 h after allergen challenge in those subjects who had previously been shown to develop a late-phase reaction. These findings complement those of a decrease in CD4 T lymphocyte numbers in the peripheral blood following allergen inhalation by atopic asthmatics [46], and together suggest that a process of selective recruitment of CD4 T lymphocytes to the lung may occur in association with the late-phase asthmatic reaction to allergen bronchial challenge. Similarly, in a study employing cutaneous allergen challenge of atopic subjects [47], activated CD4 T lymphocytes were selectively recruited during the course of the late-phase reaction.

Despite the fact that sensitive ELISA and radioimmunoassays for many lymphokines are now available, detection of lymphokine secretion in vivo is very difficult owing to their low concentrations and rapid metabolism. Furthermore, the concentrations of lymphokines in the peripheral blood and BAL fluid of asthmatics may only dimly reflect those concentrations released locally in the inflamed mucosa. One useful alternative to the direct measurement of lymphokine concentrations is the detection of the synthesis of their mRNA using the technique of in situ hybridization with lymphokine-specific cDNA probes or riboprobes. Although this is not a strictly quantitative technique, it does have the advantage that it can localise the secretion of lymphokines within cells and tissues. Using this technique it was recently demonstrated that IL-5 mRNA was elaborated by cells in the bronchial mucosa of a majority of mild asthmatics but not normal controls [48]. The amount of mRNA detected correlated broadly with the numbers of activated CD4 T lymphocytes and eosinophils in biopsies from the same subjects, providing direct evidence supporting the hypothesis that activated CD4 T lymphocytes secrete IL-5 within the asthmatic bronchial mucosa which regulates the numbers and activation status of eosinophils. In a second study using in situ hybridization, the profile of lymphokine production by cells in BAL fluid of a group of mild atopic asthmatics was compared with that from a group of normal controls [49]. The cells in the BAL fluid from asthmatics showed significantly increased expression of mRNA encoding IL-2, IL-3, IL-4, IL-5 and GM-CSF but not interferon-γ as compared to the normal controls. When T lymphocytes were isolated from the remainder of the BAL cells using immunomagnetic beads, 98% of the total expression of mRNA encoding IL-4 and IL-5 was associated with the T lymphocyte fraction. In two further studies using in situ hybridization, the cutaneous inflammatory responses to challenge with allergen in atopic subjects and tuberculin in non-atopic subjects were compared [50, 51]. Both types of response (late-phase allergic and DTH) were associated with an influx of activated CD4 T lymphocytes, but whereas mRNA molecules encoding IL-2 and interferon-γ were abundant within the tuberculin reactions, very little mRNA encoding these lymphokines was observed in the late phase allergic reactions. Conversely, mRNA encoding IL-4 and IL-5 was abundant in the late-phase allergic but not the tuberculin reactions. In effect, the profiles of lymphokine secretion in the allergic and tuberculin reactions closely paralleled those of Th2 and Th1 CD4 T lymphocytes, respectively. Furthermore, the relative numbers and types of granulocytes infiltrating these reactions reflected the different patterns of lymphokine release [52]. The detection of mRNA does not necessarily equate with protein synthesis and it will need to be shown that translation and secretion of these lymphokines also occurs. Furthermore, as discussed above, T lymphocytes are not the only potential sources of these lymphokines. Nevertheless, these observations provide direct evidence in support of the hypothesis that activated T lymphocytes, through their patterns of lymphokine secretion, regulate the types of granulocyte which participate in inflammatory reactions. Furthermore, they demonstrate that Th1 and Th2 CD4 T lymphocyte responses can be detected in humans under physiological conditions, and that the antigen specificity of the T lymphocytes might be one factor which determines which type of response is initiated.

Summary and Conclusions

Evidence is accumulating that chronic asthmatic inflammation represents a specialized form of cell-mediated immunity, in which secretion of specific lymphokines principally by activated T lymphocytes brings about local mucosal accumulation and activation of specific granulocytes, particularly eosinophils. The release of inflammatory mediators from these cells results in tissue damage and may contribute to further recruitment of inflammatory cells. This scheme does not envisage an indispensable role for antibody-dependent inflammatory mechanisms, including those mediated by IgE. There is little doubt, on the other hand, that IgE-mediated mechanisms may *exacerbate* asthma. It is possible to hypothesize that asthma results from an inherited defect in the local mucosal environment (such as a defect in epithelial integrity or a dysregulation of local immune tolerance to environmental allergens) which facilitates the development of chronic cell-mediated inflammation accompanied in some individuals by inappropriate IgE synthesis. It will be important in the future to delineate the relative importance of cell-mediated and IgE-mediated mechanisms in these diseases in order to plan effective therapeutic strategies. Glucocorticoids are the only drugs which have been shown unequivocally to reduce bronchial hyperresponsiveness in asthma. Glucocorticoids strongly inhibit T lymphocyte and monocyte function, whereas they exert little or no direct inhibitory effect on granulocytes such as mast cells and eosinophils [53]. This might be taken as evidence that suppression of cell-mediated inflammation is more effective for the therapy of these diseases than inhibition of granulocyte function. This premise is further supported by a recent double-blind trial of cyclosporin A (CsA) therapy in a group of chronic, severe glucocorticoid-dependent asthmatics [54]. CsA therapy resulted in dramatic improvements in lung function in some of these patients whose disease would previously have been considered to be 'irreversible' on glucocorticoid therapy alone. CsA potently and relatively specifically inhibits T lymphocyte function, although there is emerging evidence that it inhibits granulocytes such as mast cells at higher concentrations.

References

1 Houston JC, De Navasquez S, Trounce JR: A clinical and pathological study of fatal cases of status asthmaticus. Thorax 1953;8:207–213.
2 Dunnill MS: The pathology of asthma with special reference to changes in the bronchial mucosa. J Clin Pathol 1960;13;27–33.
3 Dunnill MS, Massarella GR, Anderson JA: A comparison of the quantitative anatomy of the bronchi in normal subjects, in status asthmaticus, in chronic bronchitis and in emphysema. Thorax 1969;24:176–179.

4 Filley WV, Holley KE, Kephart GM, Gleich GJ: Identification by immunofluorescence of eosinophil granule major basic protein in lung tissues of patients with bronchial asthma. Lancet 1982;ii: 11–16.
5 Jeffery PK, Godfrey RW, Adelroth E, Nelson F, Rogers A, Johansson S-A: Effects of treatment on airway inflammation and thickening of reticular collagen in asthma: A quantitative light and electron microscopic study. Am Rev Respir Dis 1992;145:890–899.
6 Roche WR, Beasley R, Williams JH, Holgate ST: Subepithelial fibrosis in the bronchi of asthmatics. Lancet 1989;i:520–523.
7 Azzawi M, Bradley B, Jeffery PK, Frew AJ, Wardlaw AJ, Knowles G, Assoufi B, Collins JV, Durham SR, Kay AB: Identification of activated T lymphocytes and eosinophils in bronchial biopsies in stable atopic asthma. Am Rev Respir Dis 1990;142:1410–1413.
8 Wardlaw AJ, Dunnette S, Gleich GJ, Collins JV, Kay AB: Eosinophils and mast cells in bronchoalveolar lavage in mild asthma: Relationship to bronchial hyperreactivity. Am Rev Respir Dis 1988; 137:62–69.
9 Kirby JG, Hargreave FE, Gleich GJ, O'Byrne PM: Bronchoalveolar cell profiles of asthmatic and non-asthmatic subjects. Am Rev Respir Dis 1987;136:379–383.
10 Jeffery PK, Wardlaw AJ, Nelson FC, Collins JV, Kay AB: Bronchial biopsies in asthma: an ultrastructural, quantitative study and correlation with hyperreactivity. Am Rev Respir Dis 1989;140: 1745–1753.
11 Glynn AA, Michaels L: Bronchial biopsy in chronic bronchitis and asthma. Thorax 1960;15:142–153.
12 Salvato G: Some histological changes in chronic bronchitis and asthma. Thorax 1968;23:168–172.
13 Laitinen LA, Heino M, Laitinen A, Kava T, Haahtela T: Damage of airway epithelium and bronchial reactivity in patients with asthma. Am Rev Respir Dis 1985;131:599–606.
14 Beasley R, Roche W, Roberts JA, Holgate ST: Cellular events in the bronchi in mild asthma and after bronchial provocation. Am Rev Respir Dis 1989;139:806–817.
15 Bradley BL, Azzawi M, Assoufi B, Jacobson M, Collins JV, Irani A, Schwartz LB, Durham SR, Jeffery PK, Kay AB: Eosinophils, T-lymphocytes, mast cells, neutrophils and macrophages in bronchial biopsies from atopic asthmatics: Comparison with atopic non-asthma and normal controls and relationship to bronchial hyperresponsiveness. J Allergy Clin Immunol 1991;88:661–674.
16 Fabbri LM, Danielli D, Crescioli S, Bevilaqua P, Meli S, Saetta M, Mapp CE: Fatal asthma in a subject sensitised to toluene diisocyanate. Am Rev Respir Dis 1988;137:1494–1498.
17 Bentley AM, Maestrelli P, Saetta M, Fabbri LM, Robinson DR, Bradley BL, Jeffery PK, Durham SR, Kay AB: Activated T-lymphocytes and eosinophils in the bronchial mucosa in isocyanate-induced asthma. J Allergy Clin Immunol 1992;89:821–829.
18 Bentley AM, Menz G, Storz C, Robinson DR, Bradley B, Jeffery PK, Durham SR, Kay AB: Identification of T-lymphocytes, macrophages and activated eosinophils in the bronchial mucosa in intrinsic asthma: Relationship to symptoms and bronchial responsiveness. Am Rev Respir Dis 1992;146:500–506.
19 Mattoli S, Mattoso VL, Soloperto M, Allegra L, Fasoli A: Cellular and biochemical characteristics of bronchoalveolar lavage fluid in symptomatic nonallergic asthma. J Allergy Clin Immunol 1991; 98:794–802.
20 Sanderson CJ, Warren DJ, Strath M: Identification of a lymphokine that stimulates eosinophil differentiation in vitro: Its relationship to interleukin-3 and functional properties of eosinophils produced in cultures. J Exp Med 1985;162:60–74.
21 Campbell HD, Tucker WQJ, Hort Y, Martinson ME, Mayo G, Clutterbuck EJ, Sanderson CJ, Young IG: Molecular cloning, nucleotide sequence and expression of the gene encoding human eosinophil differentiation factor (interleukin-5). Proc Natl Acad Sci USA 1987;84:6629–6633.
22 Rothenberg ME, Owen WF, Silberstein DS, Soberman RJ, Austen KF, Stevens RL: Eosinophils cocultured with endothelial cells have increased survival and functional properties. Science 1987; 237:645–647.
23 Rothenberg ME, Petersen J, Stevens RL, Silberstein DS, McKenzie DT, Austen KF, Owen WF: IL-5 dependent conversion of normodense human eosinophils to the hypodense phenotype uses 3T3 fibroblasts for enhanced viability, accelerated hypodensity and sustained antibody-dependent cytotoxicity. J Immunol 1989;143:2311–2316.

24 Wang JM, Rambaldi A, Biondi A, Chen ZG, Sanderson CJ, Mantovani A: Recombinant human interleukin-5 is a selective eosinophil chemoattractant. Eur J Immunol 1989;19:701–705.
25 Walsh GM, Hartnell A, Wardlaw AJ, Kurihara K, Sanderson CJ, Kay AB: IL-5 enhances the in vitro adhesion of human eosinophils, but not neutrophils, in a leukocyte integrin (CD11/18) dependent manner. Immunology 1990;71:258–265.
26 Lopez AF, Sanderson CJ, Gamble JR, Campbell HD, Young IG, Vadas MA: Recombinant human interleukin-5 is a selective activator of eosinophil function. J Exp Med 1988;167:219–224.
27 Rothenberg ME, Owen WF, Silberstein DS, Woods J, Soberman RJ, Austen KF, Stevens RL: Human eosinophils have prolonged survival, enhanced functional properties and become hypodense when exposed to human interleukin-3. J Clin Invest 1988;81:1986–1992.
28 Silberstein DS, Owen WF, Gasson JC, Di Persio JF, Golde DW, Bina JC, Soberman R, Austen KF, David R: Enhancement of human eosinophil cytotoxicity and leukotriene synthesis by biosynthetic (recombinant) granulocyte-macrophage colony stimulating factor. J Immunol 1986;137:3290–3294.
29 Lopez AF, Williamson DJ, Gamble JR, Begley CG, Harian JM, Klebanoff SJ, Waltersdorph A, Wong G, Clark SC, Vadas MA: Recombinant human granulocyte-macrophage colony stimulating factor stimulates in vitro mature human eosinophil and neutrophil function, surface receptor expression and survival. J Clin Invest 1986;78:1220–1228.
30 Valerius T, Repp R, Kalden JR, Platzer E: Effects of IFN on human eosinophils in comparison with other cytokines. J Immunol 1990;145:2950–2958.
31 Raghavachar A, Fleischer S, Frickhofen N, Heimpel H, Fleischer B: T-lymphocyte control of human eosinophilic granulopoiesis. J Immunol 1987;139:3753–3758.
32 Walker C, Virchow J-C, Bruijnzeel PLB, Blaser K: T cell subsets and their soluble products regulate eosinophilia in allergic and nonallergic asthma. J Immunol 1991;146:1829–1835.
33 Moqbel R, Hamid Q, Ying S, Barkans J, Hartnell A, Tsicopoulos A, Wardlaw AJ, Kay AB: Expression of mRNA and immunoreactivity for the granulocyte/macrophage colony stimulating factor (GM-CSF) in activated human eosinophils. J Exp Med 1991;174:749–752.
34 Kita H, Ohnishi T, Okubo Y, Weiler D, Abrams JS, Gleich GJ: GM-CSF and interleukin-3 release from human peripheral blood eosinophils and neutrophils. J Exp Med 1991;174:745–748.
35 Mosmann TR, Coffman RL: Th1 and Th2 cells: Different patterns of lymphokine secretion lead to different functional properties. Annu Rev Immunol 1989;7:145–173.
36 Stevens TL, Bossie A, Sanders VM, Fernandez-Botran R, Coffman RL, Mosmann TR, Vitetta ES: Subsets of antigen-specific helper T-cells regulate isotype secretion by antigen-specific B-cells. Nature 1988;334:255–258.
37 Coffman RL, Seymour BW, Lebman DA, Kiraki DD, Christiansen JA, Shrader B, Cherwinski HM, Sarelkoul HFJ, Finkelman FD, Bond MW, Mosmann TR: The role of helper T-cell products in mouse B-cell differentiation and isotype regulation. Immunol Rev 1988;102:5–28.
38 Reynolds DS, Boom WH, Abbas AK: Inhibition of B-lymphocyte activation by interferon-γ. J Immunol 1987;139:767–773.
39 Cher DJ, Mosmann TR: Two types of murine helper T-cell clone. II. Delayed-type hypersensitivity is mediated by Th1 clones. J Immunol 1987;138:3688–3694.
40 Paliard X, de Waal Malefijt R, Yssel H, Blanchard D, Chretien I, Abrams J, de Vries J, Spits H: Simultaneous production of Il-2, IL-4 and IFN-γ by activated human CD4+ and CD8+ T-cell clones. J Immunol 1988;141:849–855.
41 Corrigan CJ, Hartnell A, Kay AB: T-lymphocyte activation in acute severe asthma. Lancet 1988;i:1129–1131.
42 Corrigan CJ, Kay AB: CD4 T-lymphocyte activation in acute severe asthma. Relationship to disease severity and atopic status. Am Rev Respir Dis 1990;141:970–977.
43 Graham DR, Luksza AR, Evans CC: Bronchoalveolar lavage in asthma. Thorax 1985;40:717.
44 Walker C, Kaegi MK, Braun P, Blaser K: Activated T-cells and eosinophilia in bronchoalveolar lavages from subjects with asthma correlated with disease severity. J Allergy Clin Immunol 1991;88:935–942.
45 Metzger WJ, Zavala D, Richerson HB, Moseley P, Iwamota P, Monick M, Sjoerdsma K, Hunninghake GW: Local allergen challenge and bronchoalveolar lavage of allergic asthmatic lungs: Description of the model and local airway inflammation. Am Rev Respir Dis 1987;135:433–440.

46 Gerblich AA, Campbell AE, Schuyler MR: Changes in T-lymphocyte subpopulations after antigenic bronchial provocation in asthmatics. N Engl J Med 1984;310:1349–1352.
47 Frew AJ, Kay AB: The relationship between infiltrating CD4+ T-lymphocytes, activated eosinophils and the magnitude of the allergen-induced late phase cutaneous reaction. J Immunol 1988; 141:4158–4164.
48 Hamid Q, Azzawi M, Ying S, Moqbel R, Wardlaw AJ, Corrigan CJ, Bradley B, Durham SR, Collins JV, Jeffery PK, Quint DJ, Kay AB: Expression of mRNA for interleukin-5 in mucosal bronchial biopsies from asthma. J Clin Invest 1991;87:1541–1546.
49 Robinson DS, Hamid Q, Ying S, Tsicopoulos A, Barkans J, Bentley AM, Corrigan C, Durham SR, Kay AB: Evidence for a predominant 'Th2-type' bronchoalveolar lavage T-lymphocyte population in atopic asthma. N Engl J Med 1992;326:298–304.
50 Kay AB, Ying S, Varney V, Gaga M, Durham SR, Moqbel R, Wardlaw AJ, Hamid Q: Messenger RNA expression of the cytokine gene cluster IL-3, IL-4, IL-5 and GM-CSF in allergen-induced late phase cutaneous reactions in atopic subjects. J Exp Med 1991;173:775–778.
51 Tsicopoulos A, Hamid Q, Varney V, Ying S, Moqbel R, Durham SR, Kay AB: Preferential messenger RNA expression of Th1-type cells (IFN-γ+, IL-2+) in classical delayed-type (tuberculin) hypersensitivity reactions in human skin. J Immunol 1992;148:2058–2061.
52 Gaga M, Frew AJ, Varney VA, Kay AB: Eosinophil activation and T-lymphocyte infiltration in allergen-induced late phase skin reactions and classical delayed-type hypersensitivity. J Immunol 1991;147:816–822.
53 Schleimer RP: Effects of glucocorticosteroids on inflammatory cells relevant to their therapeutic applications in asthma. Am Rev Respir Dis 1990;141:S59–S69.
54 Alexander AG, Barnes NC, Kay AB: Cyclosporin A in corticosteroid-dependent chronic severe asthma. A randomised double-blind placebo-controlled crossover trial. Lancet 1992;339:324–328.

A.B. Kay, MD, PhD, Department of Allergy and Clinical Immunology,
National Heart and Lung Institute, Dovehouse Street, London SW3 6LY (UK)

Evidence for a Biological Activity of Anti-Tissue Antisera on an Isolated Cell System

Marie-Hélène Bobo[a], Richard Magous[a], Philippe Pouderoux[b], Jean-Louis Balmes[b], Pierre Mingard[c], Jean-Pierre Bali[a]

[a] Faculté de Pharmacie, Université de Montpellier I, Montpellier;
[b] Service d'Hépato-Gastroentérologie, CHRU Carémeau, Nîmes, France;
[c] Serolab SA, Lausanne, Switzerland

Introduction

Antibodies are able to modulate the immune response through interactions with idiotypes and with cell-bound receptors. Immunoglobulins are currently used in the treatment of medullar aplasia (antilymphocyte serum), viral, infections and auto-immune diseases (human gammaglobulins) [1]. However, little is known about the cellular regulation mediated by immune sera raised against tissue antigens [2, 3]. On the basis of some interesting clinical results obtained with such antisera [4], in vitro experiments were performed which suggested a direct cellular regulation by these antisera: horse antibodies against hog reticulo-endothelial tissue do not exhibit any species specificity and recognize human lymphocytes and macrophages [5]. Moreover, these antibodies present a weak mitogenic effect on lymphocytes in culture and, when added to stimulated lymphocytes, exert an immunodepressing effect at low concentration.

In this work, we followed a similar approach to determine whether antibodies raised against digestive tissues can modulate a biological response in a tissue from the same origin: isolated smooth muscle cells from gastric antrum [6] were used to evaluate the role of an immune serum raised against hog gastric tissues (SER 292) in the control of cellular contraction.

Materials and Methods

Isolation of Smooth Muscle Cells from Gastric Antrum

Smooth muscle cells from the gastric antrum of a rabbit were prepared by the method previously described by Moummi et al. [7]. After removing the stomach, the antral part was extensively washed with iced PBS solution (pH 7.4). Muscle tissues were sliced and incubated in medium A (132 mM NaCl, 5.4 mM KCl, 5 mM Na$_2$HPO$_4$, 1 mM NaH$_2$PO$_4$, 1.2 mM MgSO$_4$, 1 mM CaCl$_2$, 25 mM HEPES 0.2% glucose, 0.2% BSA, 0.02% phenol red, pH 7.4) containing 0.25% collagenase, 0.04% pronase, 0.01% STI and gassed with 100% O$_2$; after 60 min at 31 °C, the incubation medium was filtered, diluted with medium A and centrifuged at 150 g for 2.5 min. The cell fragments were dispersed into single cells by passages in and out the inverted wide end of a 5-ml pipette. The resulting cell suspension was filtered through a nylon mesh. Isolated cells from the two incubations diluted in medium B (Earle's balanced salt solution containing 10 mM HEPES and 0.2% BSA, pH, 7.4) were pooled and counted. Viability (estimated by trypan blue exclusion) was always greater than 90%. This protocol usually yielded about 10^7 cells (SMC) per rabbit antrum.

Measurements of Isolated Cell Contraction

Smooth muscle cell suspension (15×10^4 cells in 0.5 ml) was added to 0.05 ml of medium B containing the agents to be tested. After 30 s at 31 °C, the reaction was stopped by adding 0.05 ml of glutaraldehyde solution (final concentration: 2%). In control experiments, 0.05 ml of medium B was used instead of the agent solution. For the study with anti-F(ab')2 and anti-Fc fragments, globulin fraction obtained from the various immune sera was preincubated with these anti-IgG fragments for 18 h at +4 °C before measuring contraction. The mean cell length was statistically determined by video-microscopic measurements of 112 cells. Contractile response was expressed as the percentage of decrease in average cell length compared to the mean length of control cells.

Calcium Influx

^{45}Ca^{2+} influx was evaluated by incubating cells (5×10^5 cells in 1 ml) with 10 µCi ^{45}Ca^{2+} for various time (1–30 min) in the presence or not of stimulants. At the end of the incubation period, 0.5 ml of the cell suspension was centrifuged through an oil phase composed of vegetable oil and dibutyl phtalate (3:10 v/v). Supernatant was removed and the cell pellet was solubilized by the addition of 0.1 ml 10% perchloric acid, then the radioactivity was determined by liquid scintillation.

Clinical Study

Six selected patients (male, 22–26 years old) were treated with one suppository of SER 292 or SENI every 2 days for 105 days. Blood sampling was done every 15 days and the different human sera were tested for their ability to contract isolated gastric smooth muscle cells alone or after incubation with either antihorse anti-F(ab')2 or antihuman anti-F(ab')2 fragments. Contraction was evaluated following the above protocol.

Statistical Analysis

Results are expressed as the mean ± SEM. Significance was assessed by Student's t test, p values smaller than 0.05 being considered significant.

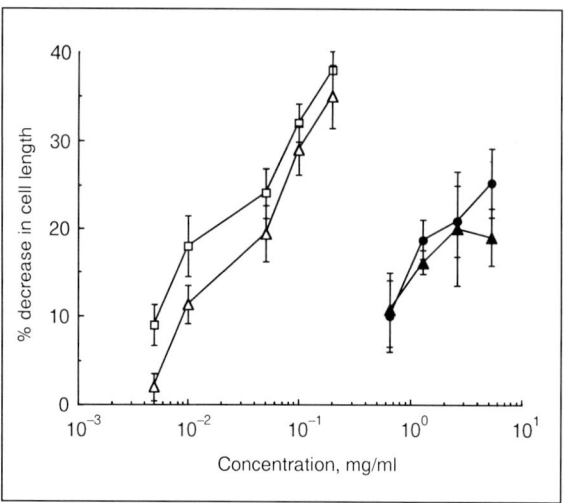

Fig. 1. Dose-response curve for SER 292- and SER 278-induced contraction. Cells (15×10^4 in 0.5 ml) are incubated in medium B with various concentrations of SER 292 (●), SER 278 (▲), SER 292-gl (□) and SER 278-gl (△) for 30 s at 31 °C. After fixation, the length of 112 cells was measured as described in 'Materials and Methods'. Each value is the mean ± SEM from 4 separate experiments. Data are expressed as the % of reduction of the mean cell length as compared to the cells in the absence of stimulant.

Results

Immune Sera-Induced Contraction of Smooth Muscle Cells

The effect of immune sera on the contractile response of isolated smooth muscle cells from the rabbit antrum has previously been investigated [6]: immune sera against tissues from the digestive tract (SER 292, anti-stomach and SER 278, anti-colon) induced contraction of isolated cells; this contraction was time-dependent with a maximal response by 30 s and dose-dependent. In contrast, immune sera raised against other tissues (heart, kidney, liver), as well as nonimmune serum (SENI) were ineffective. In order to specify the nature of the active fraction in the immune sera, albumin and globulin fractions were sequentially separed by membrane filtration (DIAFLO XM-50) and by Sephadex G-100 chromatography. Globulin fractions from SER 292 and SER 278 displayed the same dose-dependent and time-dependent contraction, with higher potency and efficacy than total serum, whereas albumin fraction was devoid of effect (fig. 1). Globulin fractions from SENI and from the other immune sera were ineffective. A further

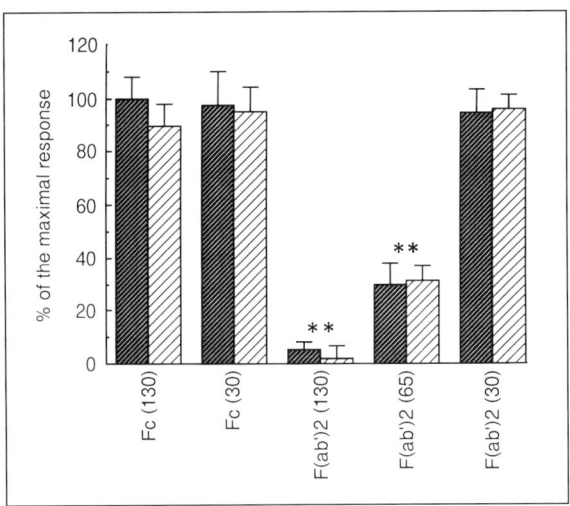

Fig. 2. Effect of anti-IgG antisera on the SER 292-induced contraction. SER 292 (☐) and SER 278 (▨) were preincubated with goat antihorse anti-Fc or goat antihorse anti-F(ab')2 for 18 h at +4 °C. Cells (15×10^4 in 0.5 ml) were incubated with various concentrations of the different complexes for 30 s at 31 °C. After fixation, the length of 112 cells was measured as described in 'Materials and Methods'. Each value was the mean ± SEM from 4 separate experiments. Data are expressed as the % of the maximal contractile response to each stimulant. ** $p < 0.01$ (Student's t test).

purification on immobilized protein G of the globulin fraction of SER 292 led to purified SER 292 IgG; this IgG fraction was shown to induce a dose-dependent contraction with higher potency than the globulin fraction.

Effect of Anti-F(ab')2 and Anti-Fc Antisera on the Contraction of SMC Induced by SER 292

In order to specify the involvement of anti-idiotype immunoglobulins in the contraction of smooth muscle cells induced by SER 292 IgG, we studied the effect of goat antihorse anti-F(ab')2 and anti-Fc antisera on the contraction induced by SER 292 IgG.

When the purified globulin fraction of SER 292 was preincubated with anti-F(ab')2, a significant reduction of the contraction was observed, which depends on the anti-F(ab')2 antiserum concentration. In contrast, the same preincubation with various concentrations of anti-Fc antiserum did not change the contractile effect observed with the globulin fraction alone (fig. 2). These results suggested that contraction was mediated by the idiotipic moiety of the immunoglobulin.

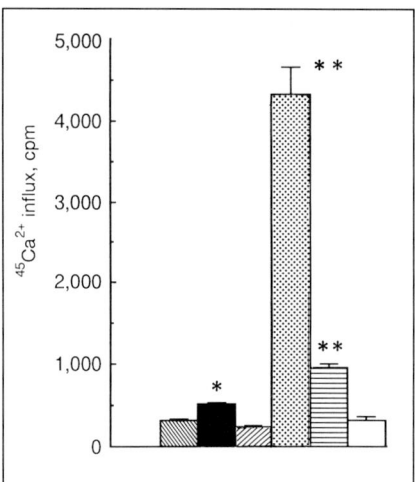

Fig. 3. $^{45}Ca^{2+}$ influx in smooth muscle cells induced by SER 292 IgG. Smooth muscle cells were incubated for 10 s at 31 °C in a medium containing $^{45}Ca^{2+}$ with the different agents: none (▨); 10 μM carbachol (■); 10 μM carbachol + 0.1 μM calcium channel blocker (▨); 32 μg/ml SER 292 IgG (▨); 32 μg/ml SER 292 IgG + 0.1 μM calcium channel blocker (▤); 32 μg/ml SENI (□). The reaction was stopped by centrifugation (1 min, 800 g) and addition of 100 μl perchloric acid (10%) to the cell pellets. $^{45}Ca^{2+}$ associated to the cell pellets (Ca^{2+} influx) was determined by scintillation counting of the solubilized pellets. ** $p < 0.01$ (Student's t test).

Role of Extracellular Calcium

Contraction of smooth muscle cells is accompanied by an increase in intracellular Ca^{2+} concentration. When the extracellular medium was depleted in Ca^{2+}, a significant reduction of the contractile effect induced by carbachol and by SER 292 IgG was observed. In addition, as evaluated by ^{45}Ca influx measurements, SER 292 IgG caused a strong rise in ^{45}Ca entry which is blocked by a calcium channel blocker (fig. 3). Both nonimmune IgG and IgG from sera raised against other tissues had no effect on calcium influx.

Clinical Study

The different human sera obtained from patients treated with SER 292 or SENI were tested for their ability to contract isolated smooth muscle cells from gastric antrum. Sera from patients treated with both SER 292 or SENI were shown to induce contraction of isolated cells. Preincubation of human serum from a patient treated with SER 292 with antihorse or antihuman Fc sera did not modify contraction due to the human serum alone. In contrast, incubation of the same human serum with antihuman anti-F(ab')2 reduced the contraction due to this human serum by about 30%, while incubation with antihorse anti-F(ab')2 reduced it by about 12%. In addition, when the purified SER 292 IgG was preincubated with antihuman anti-F(ab')2, no significant reduction of the contractile effect was observed. So, human serum from patients treated with SER 292 caused contraction of isolated cells via F(ab')2 fragments of immunoglobulins from both horse and human isotypes.

Discussion

In the present study, we examined the effect of various immune sera raised against hog digestive tissues on contraction of gastric smooth muscle cells. Our results showed that antiserum against hog gastric wall caused a time-dependent and dose-dependent contraction of cells isolated from the gastric antrum of the rabbit. In contrast, immune sera against other tissues (reticuloendothelial, heart, kidney, liver), as well as nonimmune serum, were ineffective. This contraction was shown to be due to the globulin fraction of the sera. In addition, a preincubation of this globulin fraction with goat antihorse anti-F(ab')2, but not with antihorse anti-Fc IgG fragment, reduced contraction of isolated cells, suggesting that contraction was provoked by the idiotypic moiety of the IgG.

In gastrointestinal smooth muscle cells, the rise of cytosolic free calcium concentration plays a central role in the stimulation of contraction by physiological agonists [8]. The present study confirmed the involvement of Ca^{2+} mobilization in the contraction due to SER 292 IgG as this agent caused an important increase in Ca^{2+} entry which was reduced by a calcium channel blocker, whereas both non-immune IgG and IgG from sera raised against other tissues had no effect. Shonai et al. [9] had shown that single smooth muscle cells obtained from immunized animals contract upon exposure to antigens in vitro, strongly suggesting that antigen-antibody reactions occur on the surface of the muscle cells, which in turn trigger the contraction. However, it is not known how antigen-antibody binding generates the biological signal to translate information intracellularly, leading to the development of anaphylactic contraction. Our results led us to postulate the existence of a direct interaction between the idiotypic moiety of IgG and specific components of the cell membrane which are involved in the contraction mechanism. Further studies will be necessary to determine what kind of calcium channels are involved in the mechanism of contraction induced by the purified SER 292 IgG.

Our clinical study revealed that human sera from patients treated with SER 292 caused contraction of isolated cells via F(ab')2 fragments of immunoglobulins from both horse and human isotypes. In addition, the fact that both human sera obtained after treatment with immune (SER 292) and non-immune (SENI) sera caused contraction of isolated smooth muscle cells suggests an involvement of the idiotypic network in the contractile response.

Acknowledgements

The authors are indebted to Serolab SA (Lausanne) and FST (Fondation pour la Recherche et l'Application Thérapeutique des Sérums Tissulaires) for their financial support. This work was also supported by grants from Centre National de la Recherche Scientifique, Université de Montpellier I and Institut National de la Santé et de la Recherche Médicale.

References

1 Dwyer JM: Manipulating the immune system with immunoglobulins. N Engl J Med 1992;326: 107–116.
2 Rowbothan B, Bearley RL: High dose intravenous IgG in adults with autoimmune thrombocytopenia. Lancet 1983;1:410–414.
3 Sany J, Clot J, Bonneau M, Andary M: Immunomodulating effect of human placenta-eluted gammaglobulins in rheumatoid arthritis. Arthritis Rheum 1982;25:17–21.
4 Ginsberg F: Comparative clinical evaluation of the activity of SER-316 suppository in the treatment of lumbar osteoarthrosis. Curr Med Res Opin 1991;12:413–422.
5 Clot J, Andary M: Anticorps de cheval anti-système reticulo-endothelial (SRE). Spécificité et propriétés vis à vis des cellules du SRE humain. Med Hyg 1983;41:2838–2844.
6 Ammor S, Magous R, Bali JP, Mingard P, Combepine G: Die Auswirkungen der Antigewebsimmunseren auf die Kontraktion von aus der glatten Muskulatur des Antrum pyloricum isolierten Zellen. Schweiz Med Wochenschr 1989;119:726–728.
7 Moummi C, Magous R, Strosberg D, Bali JP: Muscarinic receptors in isolated smooth muscle cells from gastric antrum. Biochem Pharmacol 1988;37:1363–1369.
8 Bitar KN, Burgess GM, Putney JW, Maklhouf GM: Source of activator calcium in isolated guinea pig and human gastric muscle cells. Am J Physiol 1986;250:G280–G286.
9 Shonai K, Okamura T, Takeuchi Y: Antigen, histamine and serotonin-induced contraction of single smooth muscle cells of guinea pig taenia coli. Jpn J Allergol 1988;37:283–293.

Marie-Hélène Bobo, Faculté de Pharmacie, Université de Montpellier I,
15 avenue Charles Flahault, F-34060 Montpellier (France)

Mechanisms of Experimental Bronchopulmonary Hyperresponsiveness as Related to Eosinophils

B. Boris Vargaftig

Unité de Pharmacologie Cellulaire, Unité Associée Institut Pasteur-INSERM 285, Paris, France

Introduction

Anaphylactic bronchoconstriction in sensitized guinea pigs is the most used test to study anti-asthmatic agents. Guinea pigs possess a developed respiratory smooth muscle, which contracts markedly in response to antigen. The resulting acute bronchoconstriction is antagonized by antihistamine agents, which are nevertheless devoid of major indication in human asthma. The anaphylactic bronchoconstriction in the guinea pig is thus essentially histamine-dependent, but may involve additional mediators according to the inflammation status of the animal. In addition to hypersensitivity, bronchial hyperreactivity, an augmented sensitivity to different stimuli and late reactions, characterises asthma. It is related to airway inflammation, particularly to the infiltration of bronchial mucosa by eosinophils and mononuclear cells [1]. To study the mechanisms of bronchial hyperreactivity, animal models are available, particularly in the guinea pig, which are based either on antigen challenge of sensitized animals, or on their exposure to irritants [2-4] or to virus [5]. Models of allergic bronchial hyperreactivity require repeated exposures to the allergen [6-8], since a single provocation with aerosolised antigen is usually not followed by bronchial hyperreactivity, despite a marked cell infiltration in the bronchoalveolar lavage fluid (BAL) [9-11]. The presence of inflammatory cells in the airways alone is thus not sufficient for inducing bronchial hyperreactivity [12].

Modulation of the establishment or of the expression of hyperresponsiveness involves the treatment of the animal before and/or together with the antigen or just before the administration of the unspecific agonist. This procedure allows to correlate bronchopulmonary hyperresponsiveness to different expressions of cell stimulation, including the BAL composition. To be completed, the in vivo approach requires some correlation between mediator formation/release and hyperresponsiveness, such as we have provided using the guinea pig isolated perfused lungs [13–18]. Indeed, the in vivo evaluation of the content in mediators of the BAL is hindered by metabolism and by difficulties in lavage collection during bronchoconstriction in presence of enhanced vascular permeability. We showed marked differences between lungs of sensitized and boosted guinea pigs, particularly an enhanced bronchoconstriction and the release of larger amounts of eicosanoids and of histamine, as compared to lungs from nonimmunized or from passively sensitized animals [14]. These modifications of the responsiveness in sensitized lungs follow the booster injection of the antigen, which operates as a microchallenge. The number of eosinophils found in the BAL fluid more than doubles in boosted guinea pigs. It is likely that repeated exposures to the allergen induce acute allergic inflammation, unperceived clinically because of the low amounts of antigen delivered.

Prophylactic anti-asthmatic drugs may prevent the development of hyperresponsiveness, even if they do not interfere with the acute manifestations of experimental asthma [17].

Eosinophils

Eosinophils are found in the bronchial submucosa of asthmatics [1] and of guinea pigs following challenge with PAF and antigen [19]. Activated eosinophils release basic proteins claimed to be toxic for the respiratory epithelium, thus exposing the submucosal structures, which are normally protected from the environment [20–22]. A rapid transvascular migration of mature cells probably accounts for early sequestration. Other mechanisms, including local differentiation of precursors and proliferation/differentiation in the bone marrow, may explain the late phase. An increased number of mature eosinophils was identified in the bone marrow of guinea pigs a few days after the booster with antigen [18]. Allergen-induced migration of eosinophils into the airways is a relevant drug target and up to now only few studies have been performed using potential inhibitors. We developed two models for in vivo eosinophil migration, using guinea pigs and mice. In both, within 24 h after challenge with antigen, the BAL population was markedly enriched in eosinophils; this was inhibited by dexamethasone, nedocromil sodium, cetirizine and mizolastine (fig. 1).

Eosinophil recruitment into the lungs requires the up-regulation of adhesive proteins at the level of the endothelium and may be followed by epithelial shed-

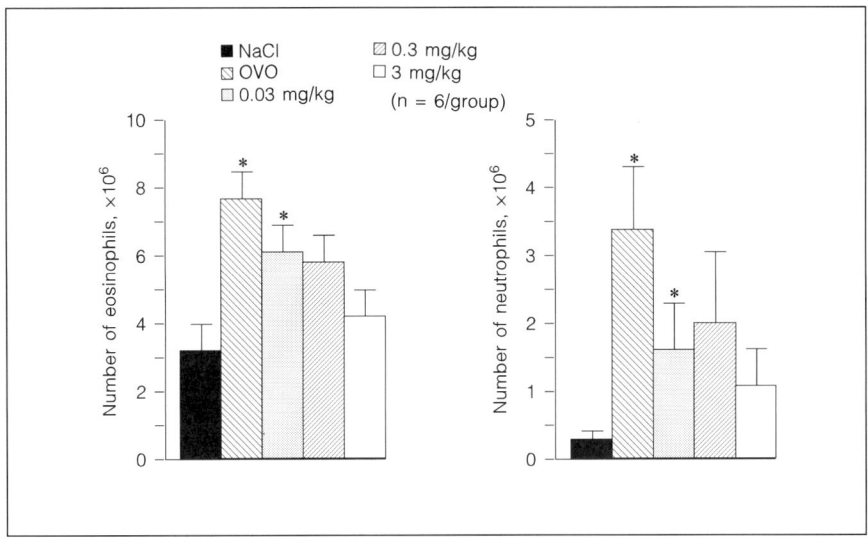

Fig. 1. Groups of 6 ovalbumin-sensitized guinea pigs were challenged with intranasal ovalbumin (0.1 mg) and the cell content of the bronchoalveolar lavage fluid was determined after 24 h. Eosinophil and neutrophil numbers are shown on top and low panels, respectively. Guinea pigs either received intranasal NaCl, to establish the basal cell content, or with ovalbumin (OVO), after the indicated doses of oral mizolastine. Eosinophil migration into the lavage fluid was dose-dependently inhibited. Significantly different (* $p < 0.05$) as compared to the basal number of cells in each case.

ding, exposure of nerve terminals and resulting bronchopulmonary hyperresponsiveness. Nevertheless, it is important to recognize that the presence of eosinophils by itself does not explain epithelial shedding or hyperresponsiveness, which are not found in humans or animals infected with parasites and showing intense hypereosinophilia. In addition, the marked eosinophil invasion into lungs and BAL following the intranasal administration of antigen to sensitized mice is not accompanied by signs of toxicity to the epithelium. Furthermore, cationic proteins other than those from eosinophils, such as poly-arginine, also induce hyperresponsiveness in rats [Coyle et al.: Am Rev Respir Dis, submitted] and guinea pigs [Desquand et al.: In preparation]. It is thus likely that, in addition to recruitment, other mechanisms are required for the full expression of the capacity of eosinophils to intensify and perpetuate asthma and allergy in general. In this context, we demonstrated that the intradermal injection of PAF to allergic humans induces a marked eosinophilic infiltration, whereas nonallergic controls show only nonspecific (neutrophil) infiltration [23, 24]. In general, sensitized guinea pigs respond more intensively to PAF or LTB$_4$ than their nonimmunized counterparts. Guinea pig peritoneal eosinophils exposed to rhIL-5 are markedly primed for migration and calcium translocation induced by PAF [25] and, finally, eosinophils collected from the BAL of sensitized and boosted guinea pigs are primed ex vivo for an enhanced migration by LTB$_4$, PAF and C5a, indicating that they have been probably activated in vitro. These results support the concept that the allergic potentiality ('atopy') is essential for the expression of the activity of the different mediators of inflammation.

Eosinophil recruitment and activation are attributed to the production of chemoattractant mediators. Silva et al. [26] demonstrated that 6–24 h after the short-lasting pleuresy which follows the intrapleural injection of PAF to rats, the number of eosinophils in the pleural cavity more than doubles. In addition, the transfer of the cell-free pleural washing from the PAF-injected cavity to that of a naive animal induced a pure eosinophilia, demonstrating that a target present in the pleural cavity of the PAF-injected rat produces an eosinophil chemoattractant. The formation of this potential mediator was suppressed by the co-injection of PAF with low amounts of its specific antagonists, with dexamethasone and nedocromil sodium. Also co-injected with PAF to the donor rat, the protein synthesis inhibitors cyclohexemide and actinomycin D suppressed eosinophilia in the donor as well as in the recipient animals. In contrast, the co-injection to a recipient animal of nedocromil sodium or of dexamethasone, of the PAF antagonists or the protein synthesis inhibitors together with the chemoattractant material generated in the PAF-injected pleural cavity, failed to interfere with eosinophil recruitment. Eosinophilia following the injection of the pleural washing generated in the donor to the recipient rat was blocked with cetirizine, which suppressed eosinophilia when administered to the donor and to the recipient rats, but failed to prevent the formation of the chemoattractant activity when injected to the donor rat together with PAF [27]. This biological activity may be accounted for by cytokine, and indeed a few of them have marked effects on eosinophils. The most selective cytokine with effects on eosinophils is interleukin 5 (IL-5) [28–31].

Conclusions

The establishment and the expression of nonspecific bronchopulmonary hyperresponsiveness in asthma is a very complex event, involving immunological loops (antigen processing, memory cells), followed by effector loops, including the interaction between preformed or neosynthesized mediators (histamine, eicosanoids, PAF) which are released rapidly, and cytokines, such as IL-5, considered to have protracted effects. We demonstrated [32] that in fact IL-5 synergizes markedly with PAF and LTB4, and has thus acute effects, not related directly to proliferation/differentiation of eosinophils. Accordingly, new potential sites of action of anti-asthmatic agents are proposed, including the interference with the formation and the effects of cytokines, or the recruitment/sequestration of eosinophils. Drugs such as nedocromil sodium, cetirizine or mizolastine belong to this new category of 'eosinophil-suppressing agents', irrespective of their other effects.

References

1 Djukanovic R, Roche WR, Wilson JW, Beasley CRW, Twentyman OP, Howarth PH, Holgate ST: Mucosal inflammation in asthma. Am Rev Respir Dis 1990;142:434–457.
2 Murlas CG, Roum JH: Sequence of pathological changes in the airway mucosa of guinea-pigs during ozone-induced bronchial hyperreactivity. Am Rev Respir Dis 1985;131:314–320.
3 Nayler RA, Mitchell HW: Airway hyperreactivity and bronchoconstriction induced by vanadate in the guinea-pig. Br J Pharmacol 1987;92:173–180.
4 Gordon T, Sheppard D, McDonald D, Di Stefano S, Scypinski L: Airway hyperresponsiveness and inflammation induced by toluene diisocyanate in guinea-pigs. Am Rev Respir Dis 1985;132:1106–1112.
5 Saban R, Dick EC, Fishleder RI, Buckner CK: Enhancement by parainfluenza 3 infection of contractile responses to substance P and capsaicin in airway smooth muscle from the guinea-pig. Am Rev Respir Dis 1987;136:586–591.
6 Ishida K, Kelly LJ, Thomson RJ, Beattie LL, Schellenberg RR: Repeated antigen challenge induces airway hyperresponsiveness with tissue eosinophilia in guinea-pigs. J Appl Physiol 1989;67:1133–1139.
7 Boichot E, Lagente V, Carré C, Waltmann P, Mencia-Huerta JM, Braquet P: Bronchial hyperresponsiveness and cellular infiltration in lungs from aerosol-sensitized and antigen-exposed guinea-pigs. Clin Exp Allergy 1991;21:67–76.
8 Kips JC, Cuvelier CA, Pauwels RA: Effect of acute and chronic antigen inhalation on airway morphology and responsiveness in actively sensitized rats: Am Rev Respir Dis 1992;145:1306–1310.
9 Hutson PA, Church MK, Clay TP, Miller P, Holgate ST: Early and late phase bronchoconstriction after allergen challenge of nonanesthetized guinea-pigs. I. The association of disordered airway physiology leukocyte infiltration. Am Rev Respir Dis 1988;137:548–557.
10 Dunn CJ, Elliot GA, Oostveen JA, Richards IM: Development of a prolonged eosinophil-rich inflammatory leukocyte infiltration in the guinea-pig asthmatic response to ovalbumin inhalation: Am Rev Respir Dis 1988;137:541–547.
11 Coyle AJ, Urwin SC, Page CP, Touvay C, Vilain B, Braquet P: The effect of the selective PAF antagonist BN 52021 on PAF- and antigen-induced bronchial hyperreactivity and eosinophil accumulation. Eur J Pharmacol 1988;148:51–58.
12 Pretolani M, Vargaftig BB: From hypersensitivity to bronchial hyperreactivity. What can we learn from studies on animal models; Biochem Pharmacol 1993; in press.
13 Pretolani M, Lefort J, Malanchère E, Vargaftig BB: Interference by novel PAF-acether antagonist WEB 2086 with the bronchopulmonary responses to PAF-acether and to active and passive anaphylactic shock in guinea-pigs. Eur J Pharmacol 1987;140:311–321.
14 Pretolani M, Lefort J, Vargafig BB: Active immunization induces lung hyperresponsiveness in the guinea pig. Am Rev Respir Dis 1988;138:1572–1578.
15 Pretolani M, Lefort J, Dumarey C, Vargaftig BB: Role of lipoxygenase metabolites for the hyperresponsiveness to platelet-activating factor of lungs from actively sensitized guinea pigs. J Pharmacol Exp Ther 1989;248:353–359.
16 Pretolani M, Lefort J, Vargaftig BB: Limited interference of specific Paf antagonists with hyperresponsiveness to Paf itself of lungs from actively sensitized guinea-pigs. Br J Pharmacol 1989;97:433–442.

17 Pretolani M, Lefort J, Silva P, Malanchère E, Dumarey C, Bachelet M, Vargaftig BB: Protection by nedocromil sodium of active immunization-induced bronchopulmonary alterations in the guinea pig. Am Rev Respir Dis 1990;141:1259–1265.
18 Pretolani M, Lefort J, Boukili MA, Bachelet CM, Vargaftig BB: Potential involvement of eosinophils and of rh interleukin-5 (IL-5) in the ex vivo lung hyperresponsiveness in the guinea pig (abstract). Am Rev Respir Dis 1991;143:A14.
19 Lellouch-Tubiana A, Lefort J, Simon MT, Pfister A, Vargaftig BB: Eosinophil recruitment into guinea pig lungs after PAF-acether and allergen administration. Am Rev Respir Dis 1988;137: 948–952.
20 Frigas E, Gleich GJ: The eosinophil and the pathophysiology of asthma. J Allergy Clin Immunol 1986;77:527–537.
21 Gleich GJ, Frigas E, Loegering DA, Wasson DL, Steinmuller D: Cytotoxic properties of the eosinophil major basic protein. J Immunol 1979;123:2925–2927.
22 Venge P: The human eosinophil in inflammation. Agents Actions 1990;29:122–126.
23 Hénocq E, Vargaftig BB: Accumulation of eosinophils in response to intracutaneous paf-acether and allergens in man. Lancet 1986;ii:1378–1379.
24 Hénocq E, Vargaftig BB: Skin eosinophilia in atopic patients. J Allergy Clin Immunol 1988;81: 691–695.
25 Coëffier E, Joseph D, Vargaftig BB: Activation by recombinant human interleukin 5 (rh-IL5) of guinea-pig eosinophils: Selective priming to Paf-acether and interference of its antagonists. J Immunol 1991;147:2295–2308.
26 Silva PMR, Martins MA, Castro-Faria Neto HC, Cordeiro RSB, Vargaftig BB: Generation of an eosinophilotactic activity in the pleural cavity of PAF-acether injected rats. J Pharmacol Exp Ther 1991;257:1039–1044.
27 Martins MA, Pasquale CP, Silva PMR, Pires ALA, Ruffie C, Cordeiro RSB, Vargaftig BB: Interference of cetirizine with the late eosinophil accumulation induced by either PAF-acether or compound 48/80. Br J Pharmacol 1992;105:176–180.
28 Sanderson CJ, Campbell HD, Young IG: Molecular and cellular biology of eosinophil differentiation factor (interleukin-5) and its effect on human and mouse B cells. Immunol Rev 1988;102: 29–50.
29 Yamaguchi Y, Suda T, Suda J, Eguchi M, Miura Y, Harada N, Tominaga A, Takatsu K: Purified interleukin-5 (IL-5) supports the terminal differentiation proliferation of murine eosinophilic precursors. J Exp Med 1988;167:43–48.
30 Chand N, Harrison JE, Rooney S, Pillar J, Jakubicki R, Nolan K, Diamantis W, Sofia RD: Anti-IL-5 monoclonal antibody inhibits allergic late phase bronchial eosinophilia in guinea-pigs: A therapeutical approach. Eur J Pharmacol 1992;211:121–123.
31 Gulbenkian AR, Egan RW, Fernandez X, Jones H, Kreutner W, Kung T, Payvandi F, Sullivan L, Zurcher JA, Watnick AS: Interleukin-5 modulates eosinophil accumulation in allergic guinea-pig lung. Am Rev Respir Dis 1992;146:263–265.
32 Pretolani M, Lefort J, Leduc D, Vargaftig BB: Effect of human recombinant interleukin-5 on in vivo lung responsiveness to PAF in actively sensitized guinea-pigs. Br J Pharmacol 1992;106:677–684.

B. Boris Vargaftig, Unité de Pharmacologie Cellulaire,
Unité Associée Institut Pasteur-INSERM 285,
25, rue du Dr. Roux, F-75015 Paris (France)

Acetaldehyde Induces a Bronchoconstrictor Response in Guinea Pigs

A Pharmacological Study

Ferruccio Berti[a], Giuseppe Rossoni[b], Alberto Buschi[a], Mariella Robuschi[a], Fabio Trento[a], Davide Della Bella[a]

[a] Department of Pharmacology, Chemotherapy, Medical Toxicology, and
[b] Institute of Pharmacological Sciences, University of Milan, Italy

The ability of ethanol to induce severe changes of bronchomotor tone in asthmatic and nonasthmatic patients has been already observed [1, 2] but the mechanism(s) responsible for this event has not been completely elucidated. The rapid decrease of specific airway conductance associated with symptoms of vasomotor sensitivity, shown by young subjects after ethanol ingestion, suggested that this compound may have acted by releasing one or more chemical mediators with bronchoactive and vasoactive properties [1]. In this respect it is important to underline that the patients under examination were of the same ethnic group (i.e. Asiatic). This population is considered at low risk of chronic alcoholism since their levels of aldehyde dehydrogenase activity (enzymes implicated in acetaldehyde oxidation) are very low due to genetic disposition [3–5].

On the other hand the inhibition of these enzymes with disulfiram represents the main pharmacological intervention for the treatment of chronic alcoholism. In fact, upon alcohol assumption the irreversible inactivation of aldehyde dehydrogenase activity caused by disulfiram brings about a marked increase in blood concentration of acetaldehyde accompanied with respiratory difficulty and spread vasodilation (acetaldehyde syndrome) [6]. All these clinical observations prompted us to investigate the ability of acetaldehyde to induce bronchoconstriction in anesthetized guinea pig, to establish the reproducibility of this phenomenon and to elucidate, through pharmacological analysis, the mode of action of this compound.

Material and Methods

Whole Animal Preparation

Male Hartley strain guinea pigs (350–400 g body weight) were anesthetized with ethyl-urethane (1.5 g/kg i.p.) and prepared as originally described by Konzett and Rössler [7] for intratracheal pressure (ITP) recording. Particularly, the trachea was cannulated for artificial ventilation which has been performed by a pump operating on a partially closed circuit (8 ml/kg, stroke volume; 70 cycles/min). To avoid spontaneous breathing the animals were treated with pancuronium bromide injected in the jugular vein at the dose of 2 mg/kg. Systemic blood pressure (BP) was also monitored via left carotid artery.

Acetaldehyde (purity >99.5%) diluted in saline at the concentrations of 2.5, 5 and 10%, was injected intravenously at the dose of 1 ml/kg.

All changes related to ITP and BP were measured by pressure transducers (HP-270 and HP-1280, respectively) and the signals displayed on a Hewlett-Packard multiple channel pen recorder (HP-7754A).

The increase in ITP values caused by various pharmacological treatments was expressed as percent of maximal overflow according to Davies and Johnston [8]. Maximal ITP values were determined at the start of each experiment by briefly clamping the trachea.

Histamine Assay

In order to determine the amount of histamine circulating in anesthetized guinea pigs, small aliquots of blood (0.5 ml) were collected from the carotid artery before and after acetaldehyde challenge (at the peak of the bronchoconstriction). Thereafter, the blood was processed according to the method described by Shore et al. [9] and the extracted histamine was estimated by a fluorimetric procedure. In our experimental conditions the recovery of histamine was superior to 95% and the basal hematic concentration of the autacoid was in the range (60.6 ± 2.5 ng/ml of blood) of that reported by other authors [10, 11].

Drugs

The following drugs were used: acetaldehyde (Merck, Darmstadt, FRG); pyrilamine maleate, atropine sulfate, indomethacin, captopril, thiorphan, substance P, histamine dihydrochloride, ethyl urethane (Sigma, St. Louis, Mo., USA); pancuronium bromide (N.V. Organon, Oss, The Nederlands); [D-Pro^4, D-Trp^{7,9}]-substance P fragment 4-11 (Peninsula Laboratories, Belmont, Calif., USA).

Data Analysis

In all experiments, differences between control and treatment groups were analysed for statistical significance using a one-way analysis of variance (ANOVA) and Student's t test (two-tailed) for paired or unpaired samples as appropriate. $p < 0.05$ was accepted as significant. In all figures, results are expressed as mean ± SEM.

Results

When acetaldehyde was injected intravenously to artificially ventilated, anesthetized guinea pigs, an increase of both ITP and systemic BP values occurred. The bronchoconstriction caused by this compound was proportional to

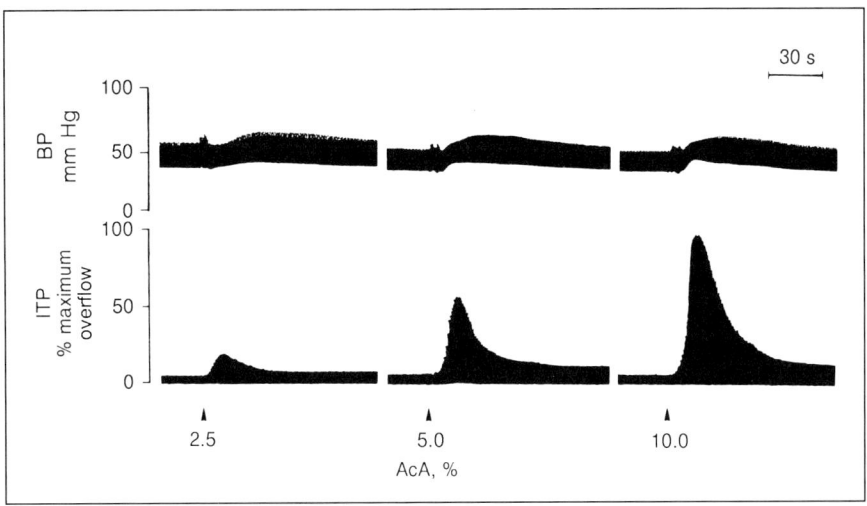

Fig. 1. Original tracing showing the dose dependency of the bronchoconstrictive effect of acetaldehyde (AcA) in anesthetized guinea pigs. AcA was injected intravenously (1 ml/kg) at the indicated concentrations.

the concentration administered and was distinguished by a rapid onset of action and decline. This particular effect was marked when acetaldehyde was given at a concentration of 10% (1 ml/kg i.v.), being the peak of ITP values equal to maximal overflow (fig. 1). No tachyphylaxis was observed as doses of the compound were given repeatedly. Furthermore, the changes in respiratory smooth muscle tone induced by acetaldehyde were also combined with a substantial and proportional increase in blood histamine, in line with the severity of the bronchoconstriction (table 1).

In order to achieve a pharmacological characterization of the acetaldehyde effect, groups of animals were treated with different substances before being challenged with acetaldehyde. The treatment of the guinea pigs with full dosage atropine (0.5 mg/kg i.v.), to block muscarinic receptors and to cut down vagal reflex discharge, did not affect the acetaldehyde action on guinea pig respiratory smooth muscles (fig. 2).

Indomethacin (1 mg/kg i.v.), given to the animal to unmask a possible involvement of eicosanoids, only slightly potentiated (11%; $p > 0.05$) the bronchoconstrictive effect of acetaldehyde (fig. 2). On the contrary, pyrilamine (2 mg/kg i.v.), at doses fully antagonizing the effect of histamine (10 µg/kg i.v.) on ITP, noticeably reduced the bronchoconstriction induced by acetaldehyde (83% inhibition; $p < 0.001$) (fig. 2).

Table 1. Acetaldehyde induces a dose-dependent bronchoconstriction associated to a proportional increase of circulating histamine in anesthetized guinea pigs

Parameters	Acetaldehyde concentrations		
	2.5%	5%	10%
ITP, % maximum overflow	20±0.9	58±2.0	95±1.2
Histamine, ng/ml blood	89±1.5	191±6.4	351±9.8

Data are mean values ± SEM of 8 experiments. In each experiment the animals were challenged with the 3 different concentrations of acetaldehyde. For histamine determination 0.5 ml of blood was collected from the carotid artery in the basal condition and at the peak of the bronchoconstriction. ITP: Intratracheal pressure; maximal overflow: 25 ± 2 cm H_2O. Histamine: Concentration differences from basal condition (60.6 ± 2.5 ng/ml blood) were highly significant ($p < 0.001$).

In order to appraise the possibility that the bronchoconstriction caused by acetaldehyde might be linked to bronchoactive tachykinins, namely substance P (SP), likely released from neuropeptide-containing capsaicin-sensitive primary afferents, a number of experiments were performed with peptidase inhibitors which are known to interfere with peptide degradation [12]. Thus, when groups of animals were treated with captopril or thiorphan at the dose (2 mg/kg i.v.) which per se did not modify baseline values of ITP but potentiated the effect of exogenous SP (5 µg/kg i.v.), the bronchoconstriction induced by acetaldehyde (concentration 2.5%, 1 ml/kg i.v.) was increased more than 4 times as compared to that obtained in control guinea pigs (fig. 3). Furthermore, when the employed concentration of acetaldehyde was augmented to 5% (1 ml/kg i.v.) the blockade of endopeptidase, obtained with captopril or thiorphan, determined an irreversible bronchoconstrictive event and all the animals died in few minutes (fig. 3).

In another set of experiments using acetaldehyde at the concentration of 2.5% (1 ml/kg i.v.), the residual bronchoconstriction induced by this compound after inhibition of both muscarinic and H_1-histamine receptors activation was studied in the presence of captopril (2 mg/kg i.v.) and in guinea pigs treated with captopril combined with [D-Pro^4, D-Trp^{7,9}]-substance P 4-11 (10 mg/kg i.v.), a specific SP-receptor antagonist [13]. The results obtained with these experiments clearly show that the treatment of the animals with atropine associated to pyrilamine reduced of the 90% ($p < 0.001$) the bronchoconstriction caused by acetaldehyde (fig. 4). The remaining bronchoconstrictive action of acetaldehyde was then greatly potentiated by captopril (30 times; $p < 0.001$) and this phenomenon

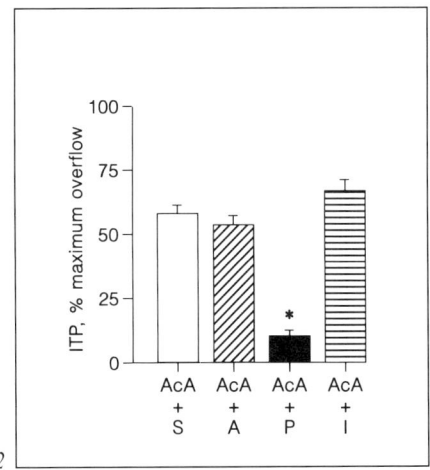
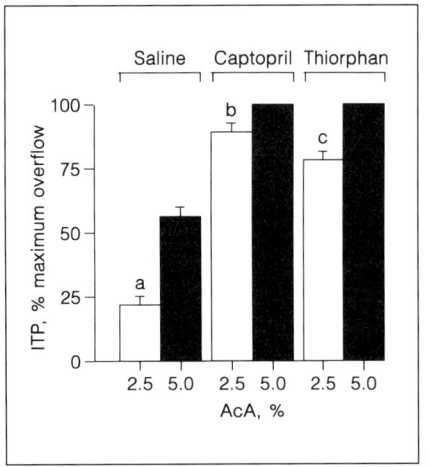

Fig. 2. Pharmacological characterization of the bronchoconstrictive effect of acetaldehyde (AcA) in anesthetized guinea pigs. Columns represent mean values and vertical bars SEM of at least 7 experiments. AcA = 1 ml/kg i.v. at the concentration of 5%; S = saline, 1 ml/kg; A = atropine, 0.5 mg/kg; P = pyrilamine, 2 mg/kg; I = indomethacin, 1 mg/kg. The antagonists were injected intravenously 3 min before AcA. ITP = intratracheal pressure; maximal overflow: 24 ± 2 cm H_2O. Significantly different from saline; * $p < 0.001$.

Fig. 3. Captopril and thiorphan injected intravenously (2 mg/kg) potentiate the bronchoconstriction induced by acetaldehyde (AcA) in anesthetized guinea pigs. Columns represent mean values and vertical bars SEM of at least 6 experiments. AcA = 1 ml/kg i.v. at the concentrations of 2.5 and 5%. In the group of animals treated with AcA at the concentration of 5% in the presence of captopril or thiorphan, the bronchospasm peaked at 100% of maximal overflow and all the animals died. The statistical differences (*a* vs. *b* and *a* vs. *c*) are highly significant ($p < 0.001$).

was prevented in a significant extent (73% inhibition; $p < 0.001$) in animals pretreated with the selective SP-receptor antagonist (fig. 4).

The possible interaction between histamine and SP in acetaldehyde-induced bronchoconstriction was also investigated. In this series of experiments the effect of SP (5 µg/kg i.v.) on ITP was not modified in 4 animals pretreated with a full dosage of pyrilamine (2 mg/kg i.v.) suggesting that at least in these experimental conditions histamine does not take part in the exogenous effect of exogenous SP on bronchial muscle tone. Similar negative results were obtained with histamine in presence of captopril or thiorphan (2 mg/kg i.v.). In fact, the bronchoconstrictive effect of histamine (10 µg/kg i.v.) was not changed in 4 animals with full inhibition of endopeptidase enzymes and blockade of SP degradation.

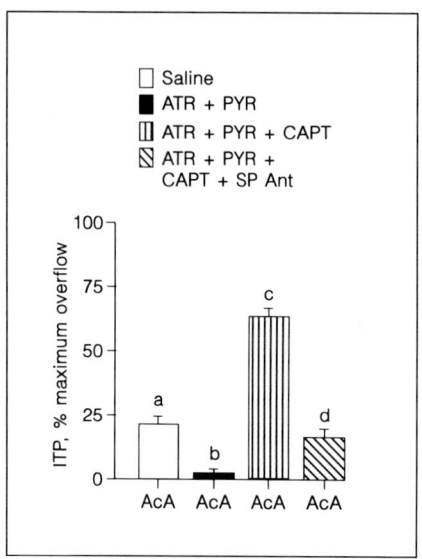

Fig. 4. Pharmacological characterization of the bronchoconstrictive effect of acetaldehyde (AcA) in anesthetized guinea pigs. Columns represent mean values and vertical bars SEM of at least 7 experiments. AcA = 1 ml/kg i.v. at the concentration of 2.5%; ATR = atropine, 0.5 mg/kg; PYR_4 = pyrilamine, 2 mg/kg; CAPT = captopril, 2 mg/kg; SP-Ant = [D-Pro^4, D-Trp^{7,9}]-Substance P 4-11, 10 mg/kg. Drugs in combination were injected intravenously 3 min before AcA. ITP = Intratracheal pressure; maximal overflow: 26 ± 1.7 cm H_2O. The statistical differences (*a* vs. *b*, *b* vs. *c* and *c* vs. *d*) are highly significant ($p < 0.001$).

Another point of interest coming afloat from these experiments was the ability of acetaldehyde, even at concentrations that did not induce changes in ITP values, to elicit hyperreactivity of guinea pig bronchial smooth muscles to SP. In fact, during acetaldehyde infusion (concentration 0.5%; 1 ml/kg/min) the bronchoconstrictive effect of SP (5 µg/kg i.v.) increased with time, being twice as potent as control responses 35 min after the start of acetaldehyde infusion. The increased responsiveness of respiratory smooth muscles to SP was a phenomenon declining with time; however, 30 min after stopping acetaldehyde infusion the effect of SP on ITP was still significantly increased (78%; $p < 0.001$) (fig. 5). The hyperresponsiveness of bronchial smooth muscles to SP induced by acetaldehyde was not significantly modified in animals treated with pyrilamine (2 mg/kg i.v.).

Furthermore, acetaldehyde infusion (0.5%; 1 ml/kg/min) did not induce hyperreactivity of the respiratory airways to both histamine (5 µg/kg i.v.) and acetylcholine (10 µg/kg i.v.).

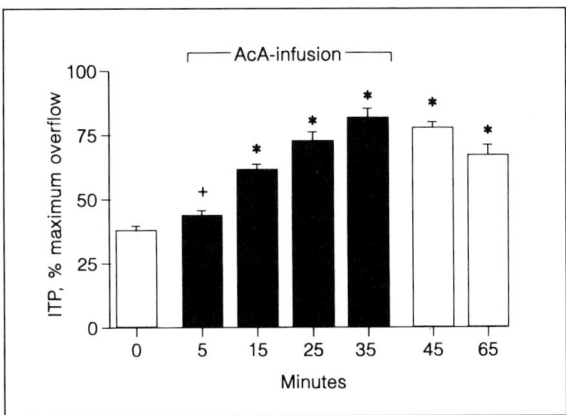

Fig. 5. Acetaldehyde (AcA) infusion potentiates the bronchoconstriction induced by SP in anesthetized guinea pigs. Columns represent mean values and vertical bars SEM of 6 experiments. AcA was infused at the concentration of 0.5% at the rate of 1 ml/kg/min. SP (5 μg/kg i.v.) at 10-min intervals did not change its activity on respiratory airways. ITP = Intratracheal pressure; maximal overflow: 26 ± 1.8 cm H_2O. + Significantly different from control ($p < 0.05$); * highly significantly different from control ($p < 0.001$).

Discussion

The present results clearly show that the intravenous injection of acetaldehyde in anesthetized guinea pigs brings about a dose-dependent increase in both intratracheal pressure and blood systemic pressure. The bronchoconstrictive effect of this compound, which is resistant to full dosage of atropine and indomethacin, does not seem to involve muscarinic receptor activation or generation of bronchoactive arachidonic acid cyclooxygenase-derived metabolites. On the contrary, since the histamine-H_1 receptor antagonist pyrilamine markedly antagonizes the bronchoconstriction induced by acetaldehyde, the great contribution made by histamine to this phenomenon may have substantial relevance. In fact, when guinea pigs are challenged with acetaldehyde, the concentration of this autacoid in the circulating blood increases and this event parallels the severity of bronchoconstriction.

In addition to the effect on airway muscles tone, the relevant enhancement in blood histamine concentration caused by acetaldehyde, should be responsible for the observed hemodynamic changes. In fact, the properties of histamine to elicit catecholamine release from adrenals and to increase generation of the vasoconstrictor thromboxane-A_2 (TXA_2) are well known. In this regard, the activation of the H_1-receptor by histamine in guinea pig perfused lungs, brings about a remarkable release in lung perfusate of TXA_2, a phenomenon which has also been reported in anesthetized animals [14, 15].

The pharmacological activity of acetaldehyde is relatively little known in animals and the data, reported by Handowsky [16] back in 1934, are not in line with our observations. In fact, this author working with anesthetized dogs showed that acetaldehyde (8–9 mg/kg i.v.) produced a stimulation of respiration with an increase in both rate and amplitude and dilation of the bronchial muscle. The discrepancy with the present findings may be due to different animal species and perhaps to the dose of acetaldehyde used. In fact, in guinea pigs the bronchoconstrictor activity of acetaldehyde is valuable with a dosage 5–10 times higher.

The alteration of bronchial tone of guinea pig observed upon acetaldehyde challenge does not appear to be a phenomenon solely related to histamine release, but the implication of mediators such as SP and other tachykinins cannot be discarded. In fact, when inhibition of endopeptidase affecting peptide degradation was obtained with captopril or thiorphan, the bronchoconstriction caused by acetaldehyde in guinea pigs was strongly potentiated: the occurrence assumed a striking picture when acetaldehyde was used at the concentration of 5% and all the animals died in a short time due to irreversible bronchial obstruction and cardiac and circulatory collapse. In this context, the possible involvement of bradykinin in acetaldehyde activity (potentiation by captopril) cannot be completely ruled out. However, since indomethacin does not significantly affect the action of acetaldehyde in guinea pig airways, and the fact that bronchoconstriction induced by bradykinin is in this animal species mediated by TXA_2 generation [15], all this restricts the possibility that bradykinin may contribute to the described effect of acetaldehyde.

The involvement of SP-like compounds in the mechanism of action of acetaldehyde is further supported by the results obtained in guinea pigs where the residual increase in bronchial resistance caused by acetaldehyde in animals pretreated with atropine and pyrilamine was conspicuously potentiated by captopril and antagonized by [D-Pro^4, D-Trp^{7,9}]-substance P 4-11, a selective NK1-receptor antagonist.

SP and related tachykinins have been implicated in various physiologic functions in mammalian species [17]. When endogenously released by stimulated afferent nerve endings in trachea and lungs these peptides induce bronchoconstriction and edema [18, 19]. In guinea pigs the acute bronchoconstrictor responses to a wide variety of irritant substances have been shown to be largely due to tachykinins release [20].

According to these reports, it is reasonable to speculate that the mode of action of acetaldehyde in guinea pigs appears rather complex and based at least on two main components: the most important one involving histamine release from pulmonary mast cells and/or basophils, and the second one concerning SP-like compounds released from capsaicin-sensitive sensory afferents in the lungs.

Furthermore, the observation that the exogenous effect of SP on bronchial tone of the guinea pigs is not affected by the H_1-histamine receptor antagonist pyrilamine seems to lessen the relevance of the contribution of histamine released by SP to the overall effect of acetaldehyde on respiratory smooth muscles. However, the failure of pyrilamine to interfere with the action of SP in airway tone upon intravenous injection may be explained by the predominance of the direct effects of this neuropeptide [21, 22].

Another point of interest arising from the present experiments is the ability of acetaldehyde to elicit in guinea pigs hyperresponsiveness of respiratory smooth muscles to SP. The phenomenon which is reversible is difficult to understand, it seems to be specific for SP, and begins at concentrations of acetaldehyde that are lower than those required for bronchoconstriction.

In conclusion, the capacity of acetaldehyde to induce bronchoconstriction in guinea pigs has been analyzed and characterized on pharmacological grounds. Even if the mode of action of this compound is not at the moment fully elucidated, the occurrence of this phenomenon may have some clinical relevance in alcoholic subjects. In fact, it has been reported that acetaldehyde dehydrogenase activity in liver biopsy specimens from noncirrhotic alcoholics is significantly reduced as compared to controls [23]. In these subjects, particularly those under treatment with angiotensin-converting enzyme inhibitors for cardiovascular disorders, a risk of bronchoconstriction upon alcohol assumption should be taken into consideration. In addition, an increase in both the number and the length of SP immunoreactive nerve fibers in airways from subjects with asthma as compared to airways from subjects without asthma has been shown [24].

Conclusion

The effect of acetaldehyde in artificially ventilated anesthetized guinea pigs was studied. The compound injected intravenously at various concentrations elicits an acute increase in intratracheal pressure combined with a proportional increment in circulating histamine. The phenomenon which was dependent on the concentration used was not significantly affected by atropine (0.5 mg/kg i.v.) or indomethacin (1 mg/kg i.v.) but was very sensitive to pyrilamine (2 mg/kg i.v.). In fact, the bronchoconstrictor response to acetaldehyde (5%, 1 ml/kg i.v.) was inhibited 83% ($p < 0.001$) by the histamine antagonist.

In guinea pigs treated with endopeptidase inhibitors, captopril or thiorphan (2 mg/kg i.v.), the effect of acetaldehyde (2.5%, 1 ml/kg i.v.) on the airways was greatly potentiated (4 times; $p < 0.001$). In this set of experiments, using acetaldehyde at the concentration of 5% (1 ml/kg i.v.) all the animals died in a few minutes due to irreversible bronchoconstriction.

The residual bronchoconstrictor effect of acetaldehyde (2.5%, 1 ml/kg i.v.) in guinea pigs treated with both atropine and pyrilamine was markedly potentiated (20 times; $p < 0.001$) by captopril (2 mg/kg i.v.) and almost fully antagonized by the specific SP-receptor antagonist [D-Pro^4, D-Trp^{7,9}]-substance 4-11 (10 mg/kg i.v.).

In conclusion, the present results clearly show that acetaldehyde induces a dose-dependent bronchoconstriction in anesthetized guinea pigs. The phenomenon which appears rather complex may be based on at least two components: the most important one involving histamine release likely from mast cells and the second one related to SP and perhaps other neurotachykinins from sensory afferent terminals of the lungs. The possibility that the released autacoids may interact in determining the bronchoconstrictor response to acetaldehyde in guinea pigs is also discussed.

References

1 Geppert EF, Boushey HA: An investigation of the mechanism of ethanol-induced bronchoconstriction. Am Rev Respir Dis 1978;118:135–139.
2 Gong Jr H, Tashkin DP, Calvarese BM: Alcohol-induced bronchospasm in an asthmatic patient. Chest 1981;80:167–173.
3 Ewing JA, Rouse BA, Pellizzari ED: Alcohol sensitivity and ethnic background. Am J Psychiatry 1974;131:206–210.
4 Wolff PH: Ethnic differences in alcohol sensitivity. Science 1972;175:449–450.
5 Yoshida A, Huang IY, Ikawa M: Molecular abnormality of an inactive aldehyde dehydrogenase variant commonly found in Orientals. Proc Natl Acad Sci USA 1984;81:258–261.
6 Kitson TM: The disulfiram-ethanol reaction. J Stud Alcohol 1977;38:96–113.
7 Konzett H, Rössler R: Versuchsanordnung zu Untersuchungen an der Bronchialmuskulatur. Naunyn Schmiedebergs Arch Exp Pathol Pharmakol 1940;195:556–567.
8 Davies GE, Johnston TP: Quantitative study on anaphylaxis in guinea-pig passively sensitized with homologous antibody. Int Arch Allergy Appl Immunol 1971;41:648–654.
9 Shore PA, Burkhalter A, Cohn VH: A method for the fluorometric assay of histamine in tissue. J Pharmacol Exp Ther 1959;127:182–186.
10 Beall GN: Histamine in guinea pig tissues. Arch Int Pharmacodyn 1966;159:484–495.
11 Anton AH, Sayre DF: A modified fluorometric procedure for tissue histamine and its distribution in various animals. J Pharmacol Exp Ther 1969;166:285–292.
12 Sekizawa K, Tamaoki J, Graf PD, Basbaum CB, Borson DB, Nadel JA: Enkephalinase inhibitor potentiates mammalian tachykinin-induced contraction in ferret trachea. J Pharmacol Exp Ther 1987;243:1211–1217.
13 Caranikas S, Mizrahi J, D'Orleans-Juste P, Regoli D: Antagonist of substance P. Eur J Pharmacol 1982;77:205–206.
14 Berti F, Folco GC, Nicosia S, Omini C, Pasargiklian R: The role of H_1- and H_2-receptors in the generation of thromboxane A_2 in perfused guinea-pig lungs. Br J Pharmacol 1979;65:629–633.
15 Rossoni G, Omini C, Viganò T, Mandelli V, Folco GC, Berti F: Bronchoconstriction by histamine and bradykinin in guinea-pigs: Relationship to thromboxane-A_2 generation and the effect of aspirin. Prostaglandins 1980;20:547–557.
16 Handowsky H: Au sujet de propriétés biologiques et pharmacodynamiques de l'acétaldéhyde. CR Soc Biol (Paris) 1934;117:238–241.

17   Emson PC, Diez-guerra FJ, Arai A: Mammalian tachykinins: neurochemistry and pharmacology; in Turner AJ (ed): Neuropeptides and Their Peptidases. Chichester, Ellis Horwood, 1987, pp 87–106.
18   Maggi CA: Tachykinin receptors in the airways and lung: What should we block? Pharmacol Res 1990;22:527–540.
19   Maggi CA, Giuliani S, Ballati L, Lecci A, Manzini S, Patacchini R, Ranzetti AR, Rovero P, Quartara L, Giachetti A: In vivo evidence for tachykininergic transmission using a new NK-2 receptor-selective antagonist, MEN 10,376. J Pharmacol Exp Ther 1991;257:1172–1178.
20   Lundberg JM, Saria A: Polypeptide containing neurons in airway smooth muscle. Ann Rev Pharmacol 1986;49:557–572.
21   Szolcsanyi J, Bartho L: Capsaicin-sensitive non-cholinergic excitatory innervation of the guinea-pig tracheobronchial smooth muscle. Neurosci Lett 1982;34:247–251.
22   Grunstein MM, Tanaka DT, Grunstein JS: Mechanism of substance P-induced bronchoconstriction in maturing rabbit. J Appl Physiol 1984;57:1238–1246.
23   Jenkins WJ, Peters TJ: Selectively reduced hepatic acetaldehyde dehydrogenase in alcoholics. Lancet 1980;ii:628–629.
24   Ollerenshaw SL, Jarvis D, Sullivan CE, Woolcock AJ: Substance P immunoreactive nerves in airways from asthmatics and nonasthmatics. Eur Resp J 1991;4:673–682.

Prof. Ferruccio Berti, Department of Pharmacology,
Chemotherapy and Medical Toxicology, Via Vanvitelli 32, I–20129 Milan (Italy)

Physical and Chronic Idiopathic Urticaria

Lennart Juhlin

Department of Dermatology, University Hospital, Uppsala, Sweden

Introduction

Urticaria is a limited elevated swelling of the dermis often surrounded by a reddening caused by a neurogenous reflex. The lesion is at least initially itching. A diagnosis of urticaria is as a rule easy to establish since the same reaction is seen after contact with nettles. If a single lesion persists for more than 24 h vasculitis should be considered. In chronic urticaria new wheals appear in the skin several days a week for more than 8 weeks and sometimes for many years. Angioedema (Quincke's edema, angioneurotic edema) is a less acute deeper swelling in the subcutaneous tissue often around the eyes and mouth. It is not itching but tingling and persists for 12–72 hours. The disorder can appear at the same time and have the same etiology except for hereditary angioedema. The physical urticarias are caused by factors such as pressure, heat, cold or UV irradiation. Before a diagnosis of chronic idiopathic urticaria is made, a careful case history is important to exclude the physical urticarias and to rule out that the hives are not caused by drugs or certain foods.

Mediators Involved in the Pathogenesis

Histamine is the most important mediator for inducing acute urticaria and the immediate reactions of physical urticaria. It is formed in the fine blood vessels and the mast cells which have receptors for immunoglobulin E making them sensitive to various antigens. Other mediators which can be involved especially in chronic urticaria are PAF (platelet-activating factor), kallikrein which liberates kinins, and vasoactive intestinal polypeptide (VIP) [1–3]. The mast cells can also be activated by factors such as complement or toxic substances liberated from insect bite infections or histamine-releasing drugs. The lesions in chronic urticaria often contain an increased number of eosinophils [4, 5] and the leakage of eosinophilic cationic protein might enhance the reaction. In cholinergic and adrenergic urticaria acetylcholine and adrenaline are also involved. Why suddenly someone starts to develop hives after exposure to cold, heat, irradiation or pressure is not known. An infection some weeks earlier is often reported which might have changed the immune response.

Occurrence

Acute urticaria is the most common type and is seen in 55% of our urticaria patients in the Uppsala clinic. In general practice it is much more common. Chronic urticaria persisting for more than 2 months accounts for 27% and the rest (18%) are caused by physical factors. The physical urticarias are often overlooked by the referring physician. By taking a careful case history they are usually easy to suspect and should be ruled out before making a diagnosis of chronic urticaria. Mixtures of the different types of physical urticaria can occur and are sometimes difficult to separate. The most common is urticaria factitia (8%), followed by cholinergic urticaria (4%), cold urticaria (3%), and various types of urticaria (3%).

Physical Urticarias

Urticaria factitia (= Dermographism)

Diagnosis. The patients react with an itchy localized swelling and reddening of the skin within a few minutes when a blunt object such as a pen or a nail is drawn on the skin with moderate pressure. The reaction persists for about 30 min. In some cases the response can be negative on the arms but positive on the back. Depending on the pressure used the border between normal and abnormal is not always distinct. The patients are not aware that the cause of their urticaria is scratching or that it appears on areas where there has been pressure such as sitting or leaning against a chair.

Cause. Why they suddenly start to react to pressure is often not known. Infestations with scabies, various infections and penicillin treatment are sometimes noted at the start. Dermographism also often occurs in connection with acute urticaria.

Treatment. A classic antihistamine, especially hydroxyzine, has for many years been the drug of choice. The new less or nonsedating antihistamines (table 1) are now preferred. Since dry skin is often the cause of itching and scratching, a lubricating cream can be of value to decrease the habit.

Many patients have noted that they are better in summertime when they have been sunbathing. Irradiation with UV-B decreases the sensitivity of the mast cells and can therefore be an additional treatment.

Table 1. Less or nonsedating antihistamines available in many countries

Generic name	Common trade name	Daily dose mg	Elimination half-life	route
Terfenadine	Teldane Seldane	120	23 h	urine
Astemizole	Hismanal	10	18–30 days	feces
Loratadine	Clarityne	10	10–24 h	feces, urine
Cetirizine	Zyrtec Virlix	10	10 h	urine
Acrivastine	Semprex	8 × 3	3–6 h	urine

Delayed Dermographism

This is a rare variant of dermographism which is easy to miss since the whealing and erythema first appears after 4–20 h. The response to the antihistamines is poor suggesting that other mediators are involved. Delayed dermographism can occur together with pressure urticaria and it is possible that it has a similar pathogenesis.

Pressure Urticaria

Diagnosis. After pressure on the skin the patients develop after 3–12 h a painful edema deep in the skin [2]. The most common localizations are the soles after a walk, the palms after working in the garden, the shoulders after carrying a sack or the gluteal region after sitting too long on a hard chair. Pain in the larger joints often occurs simultaneously. When the swellings are marked there is often fever, leukocytosis followed by an elevation of the sedimentation rate.

In the skin an increased infiltration of eosinophil leukocytes is the rule. The eosinophils in the blood are not increased but the eosinophilic cationic protein (ECP) in serum is increased which indicates that the eosinophilic leukocytes are involved in the pathogenesis [5, 6]. The disease is often combined with chronic urticaria and a few families with a hereditary background have been described. To provoke the lesions we let the patient walk around for an hour with two 5-kg sacks hanging over the shoulder.

Treatment. The skin lesions do usually not respond to the antihistamines even in high doses which can inhibit a concurrent chronic urticaria. From Greece there have been reports, however, that the high doses of the antihistamine cetirizine has been effective [7]. Prednisolone can in daily doses of 40 mg have an effect but prolonged treatment with this drug at high doses is not recommended. This rare form of urticaria is usually difficult to treat and persists for several years.

Cholinergic Urticaria

This urticaria is often also called heat urticaria since it appears when the body temperature is increased by physical effort, a hot bath or fever. In some patients emotional stress or spicy food can induce the typical symptoms. The typical lesions are small intensively itching papular wheals (1–4 mm diameter) surrounded by an erythema. They usually appear first on the upper part of the body and last for 10–30 min. In some cases itching is the only symptom. The disorder is most common between 15 and 30 years of age. Other symptoms reported simultaneously are headache, wheezing, abdominal pain, vomiting and diarrhea which can all be signs of cholinergic stimulation.

Treatment. Try to avoid physical effort and other known precipitating factors. Nonsedating antihistamines give temporary relief and can be recommended before jogging and sports. The symptoms usually remain for some years but decrease with increasing age.

Localized Heat Urticaria

This is a rare disorder where urticaria appears within some minutes exactly at the site where a 37–40 °C warm object has been in contact with the skin for some minutes. The patients have no systemic symptoms.

Aquagenic Urticaria

This urticaria looks clinically like cholinergic urticaria and has small itching wheals [2, 8]. It can be induced by a shower, a bath or sweating and is mainly localized to the neck and upper chest. It is assumed that an unknown water-soluble substance on the skin surface is resorbed. The patients are often females aged 7–40 years. The disorder remains for several years. Antihistamines decrease the reaction and should be tried.

Adrenergic Urticaria

The disorders can be said to be a counterpart to cholinergic urticaria. In stress situations the patient can experience a small itching urticarial papule surrounded by a typical pale halo. The symptoms can be induced by adrenaline but not by acetylcholine. Treatment with a beta-blocker like 10–20 mg propranolol twice daily is effective. If this type of urticaria is treated with adrenalin there is a risk of anaphylactic shock [9].

Cold Urticaria

After being out in cold and wet weather the exposed face or hands become red and swollen. After swimming typical wheals can appear on the body also at temperatures of 24 °C. In women who wear skirts the urticaria can be seen on the upper legs. The itching and the wheals appear especially when the patients, after being out, enter a warm room. The wheals usually disappear within 30 min.

Diagnostic Tests. (1) Ice cube test: An ice cube wrapped in a plastic foil is held on the skin for 1, 2, 4, 8 and 12 min. Thereafter the skin is warmed under a lamp for some minutes and in a positive test the urticaria appears where the ice has been applied. It is thus possible to grade the sensitivity of the urticaria and the effect of the treatment by recording the number of minutes needed for a reaction.

(2) Water test: If the patient has a history of possible cold urticaria but the ice cube test is negative immersion in cold water with floating ice cubes for 10–15 min can be performed. After warming typical urticarial wheals appear on the arm.

(3) Cold air test: If the above tests both are negative a provocation test with cold air is recommended. In a cold climate one can simply let the patient walk out on a cold and rainy day to demonstrate the lesions. The reason why the ice cube and cold water tests were negative but the cold air test was positive can be that the lowering of the temperature must be optimal. In other patients a lowering of the body temperature by sitting in a cold room (+4 °C) for 10 minutes is needed.

Temperature. Arctic clothing and good protection is essential as well as a heated car for transportation. The nonsedating antihistamines are worth trying. They can increase the time the patients can be out without getting urticaria especially if they only react to a 9- to 12-min ice cube test [6, 7]. The antihistamines should be taken at least 2 h before the patient goes out.

Desensitization. A partial refractory state develops after exposure to the cold. It can be used to decrease the sensitivity by washing for example the face with cold water every morning. There might be an initial irritation but if repeated daily the patient can tolerate subsequent cold exposure better.

Information. The patient should be informed that cold urticaria is one of the most common causes of death from drowning after swimming due to anaphylactic shock. When swimming they should know their tolerance and not be alone. They should also be careful when going on long skiing tours and know that they are more vulnerable than others when the weather turns bad. They should also be informed that in the case of accident the infusion solutions must be warmed up before use. Cold drinks can sometimes produce swelling of the throat and stomach ache. Patients with atopic dermatitis more often have cold urticaria than others. Why healthy people suddenly become sensitive to cold is not known but a preceding virus infection is often recorded which might have changed the immune system. The patients have a factor in the serum which can be passively transferred in 50% of the cases. By injecting their serum intradermally in a healthy subject the treated area will react with urticaria on exposure to cold. The patients are interested in the prognosis. The sensitivity of the ice cube test is here of little help. In some patients the sensitivity to cold disappears within a few months, in others it persists for several years especially in children. After 60 years of age the disease is rare.

Solar Urticaria

Light or solar urticaria is a rare form of urticaria which is provoked by irradiation of varying wavelengths. The most common are UV-B (300–320 nm) and/or UV-A (320–400 nm). We do not know why someone aged 20–40 years who has always enjoyed sunbathing suddenly reacts with urticaria. The swelling and reddening of the skin is seen some minutes after exposure. Areas which are exposed repeatedly tolerate the irradiation better. Systemic reactions such as headache and drop in blood pressure are common. Without treatment the patients must stay indoors during the day.

Treatment. Desensitizing by daily irradiation with the wavelengths to which the patient has reacted is the treatment of choice. One starts with a few seconds of irradiation and as soon as the reaction has disappeared one should repeat it with an increased dose and area [2, 10]. The patient can finally be irradiated on the whole body every morning for 5–20 min. If the procedure is interrupted the effect is lost immediately. Antihistamines have usually no or little effect. Plasmapheresis can also help for some months by removing a serum factor [11].

Chronic Urticaria

Occurrence

In chronic urticaria new wheals appear in the skin almost every day for more than 6–8 weeks. The symptoms can persist for several years (fig. 1). Chronic urticaria is mainly seen between 20 and 60 years of age and is somewhat more common in women than in men. It is estimated that in a population of 10 million people there are 5,000 patients with chronic uticaria.

Pathogenesis

Patients with chronic urticaria show after 5–10 h a delayed reaction to their own (autologous) serum [1, 3, 12], as well as to mediators such as kallikrein, histamine prostaglandins, and PAF [1–3]. It is possible that the mediators cause an initial leakage of a plasma factor which then causes a delayed reaction. In the delayed reactions induced by the mediators and increase of eosinophil leukocytes can be demonstrated. The eosinophils contain proinflammatory cationic proteins which might augment the reaction [5]. The presence of activated eosinophils in the skin has also been found to be increased in urticarial lesions and especially in those who had a dense perivascular infiltrate [4, 5]. In these immunohistochemical studies one used the monoclonal antibody EG-2 which recognizes both the eosinophilic cationic protein (ECP) and eosinophilic protein (= eosinophil-derived neurotoxin-EPX/EDN).

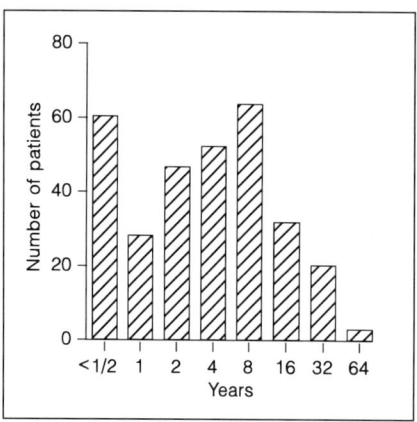

Fig. 1. Duration of urticaria in 300 patients with chronic urticaria.

An increase of eosinophils in the skin has previously been described after PAF and antigen injection in sensitized patients [13–15]. We found 10–40 eosinophils per section in the 5-hour PAF test in patients with chronic urticaria [16]. In healthy subjects PAF also induced eosinophils but only after higher doses. We found no correlation between the size of the wheal at 5 h and the number of eosinophils which conforms with the finding after testing with antigens [14] and shows that the eosinophils are at least not solely responsible for the delayed reactions but that other factors must also be involved. Treatment with the antihistamines dexchlorpheniramine and loratidine did not affect the influx of eosinophils either in the patients or in the controls [16].

Patients with chronic urticaria usually have normal eosinophil blood counts, whereas the serum ECP is increased in 45% of the patients (table 2). That serum ECP is increased but not the circulating eosinophils implies that ECP is formed by eosinophils located somewhere else in the body. The patients with increased ECP all had daily eruptions and tended to have the most severe urticaria. In the other types of urticaria studied no increase of EG-2-positive cells or ECP was found with the exception of delayed pressure urticaria (table 2).

The presence of another cationic protein found both in eosinophils and basophils, the major basic protein (MBP), has also been demonstrated in the lesions from 45% of the patients with chronic urticaria and in most patients with pressure urticaria as well as in 4 patients with solar urticaria [6, 17, 18]. For the visualization of MBP a polyclonal antibody was used which was also found dispersed on the connective tissue fibers in the dermis. Thus, the location of MBP in the dermis differs from that of ECP which is almost exclusively found in the eosinophils.

Table 2. Chronic urticaria

Type of urticaria	EG-2-positive cells (mean of 3 sections)				ECP serum, µg/l		Eosinophils/mm^3	
	0–3	4–10	11–30	>30	0–16	>16	0–470	>470
Chronic	8	3	2	2	11	9	20	–
Pressure	–	–	–	4	2	2	4	–
Vasculitic	–	1	2	2	0	4	2	2
Cold	5	–	–	–	5	–	5	–
Dermographic	5	–	–	–	5	–	4	–
Cholinergic	3	–	–	–	3	–	3	–
Healthy subjects	20	–	–	–	20	–	20	–

Figures indicate numbers of patients.

We have recently studied the location of MBP in lesions from patients with chronic urticaria using a monoclonal antibody BMK-13 which was kindly supplied by Dr. R. Moqbel in London. When we compared the results of BMK-45 with EG-2 using a double staining technique on the same section the two proteins were often found in different areas of the same eosinophils. Some eosinophils expressed only ECP and others only MBP. Further studies are needed to find out why sometimes EG-2 dominates and sometimes BMK-13.

There are several other possible mechanisms for the appearance of wheals [2]. A lack of inhibitor to the kallikrein-kinin system has been proposed [1]. Endotoxins and chemicals in food and drugs such as aspirin can increase the reactivity of mast cells and vessels. A reduced diaminase activity decreasing the breakdown of histamine could also be involved [19]. The delayed reactions in the skin can be inhibited by some antihistamines but not by all [16 for refs]. Since mediators other than histamine often seem to be involved in urticaria it can be of value to try another antihistamine if one does not work.

Investigations at First Visit

Drugs and Food. List all the drugs the patient has taken. Try to stop all antiinflammatory drugs including aspirin. Use different commercial names when asking. If a pain reliever is needed recommend paracetamol. Stop all other drugs unless they are absolutely necessary.

Since certain food additives such as benzoates and azodyes can cross-react with the antiinflammatory drugs the patient should be informed about how to avoid them for some weeks since they might trigger a reaction. Ask for special

Table 3. Butylated hydroxyanisol and toluene (BHT/BHA)

Ready-made food, bread, flakes, cakes
Milk and potato powder, dry yeast
Fats, oils, margarine, mayonnaise, dressings
Jam, marmelade, chocolate, sweets
Nonalcoholic drinks, instant drinks

---

habits of eating or drinking. Does the patient often consume licorice, menthol tablets or drink tonic water which contains quinine?

Anamnesis. Take a careful medical history. Are there any clinical signs of thyroid disease? If the patient is already taking thyroid drugs a small adjustment of the dose can improve the urticaria. Are there any clinical signs of infection? To eliminate an infection focus by surgery or antibiotics can have a favorable effect [20]. Systemic symptoms such as gastrointestinal problems, allergic manifestations, joint problems or depression are often reported and might need attention [21].

Laboratory Tests. Take a routine blood test including sedimentation rate, hemoglobin, transaminases and leukocytes with a differential count. If there are signs of eosinophils try to find the cause. Have a urine test for protein, sugar and sediment.

Drug Treatment. Prescribe an H_1 antihistamine and ask the patient to note the frequency and severity of the urticaria daily. For daytime use a less or non-sedating antihistamine is best. The patient should find out if there is any sedation before driving a car or handling machinery which could cause an accident at work.

Second Visit after 2–4 Weeks

Find out the effect of the given recommendations and treatment. If there was no or little improvement a diet without yeast is recommended and treatment with nystatin for a week. Inform the patient about the two food additives butylhydroxytoluene (BHT) and butylhydroxyanisole (BHA) which are phenols used to prevent fats from deteriorating. Give the patient a list where these compounds are common (table 3). Some patients have a problem keeping to a diet where several foods should be avoided and here one has to try the diets successively. Ask again if the patient does not react to seafood, nuts, peas or soy. If uncertain, a radioallergosorbent test (RAST) can be of value. This is of special importance for soy which can be hidden in several foods including ham and chicken if the animals have been fed on soy.

Third and Further Visits

Review the effects of given regimens. Continue with a diet if there is a clear improvement but stop it if there is no improvement. Oral provocation tests can be worth doing if one is uncertain. If the urticaria continues an antihistamine should be used regularly. The effect of changing the type of nonsedative antihistamine used should be explored. Only rarely and mainly in patients with an increased body weight can it be worth trying to increase the dose somewhat above the recommended. In patients who are very tense a weak dose of a β-blocker might be worth trying and in more flegmatic women a β-stimulator. In a very limited number of patients has an improvement been noted when combining an H_1 and H_2 receptor antagonist [22 for refs]. Cases have also been described where the addition of a calcium channel blocker like nifedipine (10–20 mg × 3) was found to be of value [22].

Other Treatments

After UV-B irradiation or UV-A plus oral psoralen good results have been described [22 for refs]. The rationale behind the treatment is that the irradiation interferes with the release of histamine from the mast cells. However, long-term risks of UV radiation need to be weighed against potential benefits.

The addition of corticosteroids to the antihistamines is often effective. If they are needed it is important to find the lowest possible effective dose. We usually start with 40 mg of prednisolone for 2 days and decrease the dose by 5 mg/day until 10 mg and then use an alternate-day regimen for 2 weeks. The prednisolone is then decreased and taken only on certain days per week. Recurrence is common, but if the response to corticosteroids is dramatic, it can be reinstituted for a short time during severe exacerbations. Because of its many adverse effects, prolonged corticosteroid treatment should not be given for chronic urticaria unless it is associated with other disorders with indications for its use.

Heparin and warfarin have been reported to inhibit urticaria and we have been able to confirm this in a few patients who received the drugs for other indications [23]. Because of the risk for internal bleedings, the treatment should only be tried under exceptional circumstances. The same holds true for cyclosporin which has been shown to temporarily inhibit the urticaria but it cannot be recommended because of its side effects [24].

The best results are obtained when the doctor and the patient have the courage and interest to get rid of the provoking factors and try to be better than the natural tendency to heal. As long as the wheals continue the patient should come back at regular intervals to try to find the cause and best treatment. We should not give up before the wheals do it.

References

1 Juhlin L, Michaëlsson G: Cutaneous reaction to kallikrein, bradykinin and histamine in healthy subjects and in patients with urticaria. Acta Derm Venereol (Stockh) 1969;49:26–36.
2 Czarnetski BM: Urticaria. Berlin, Springer, 1986.
3 Juhlin L, Rihoux J-P: Effect of cetirizine on cutaneous reactions to PAF, kallikrein and serum in patients with chronic urticaria. Acta Derm Venereol (Stockh) 1990;70:151–153.
4 Tai P-C, Spry JF, Peterson C, Venge P, Olsson I: Monoclonal antibodies distinguish between storage and secreted forms of eosinophilic cationic protein. Nature 1984;309:182–184.
5 Juhlin L, Venge P: Eosinophilic cationic protein (ECP) in skin disorders. Acta Derm Venereol (Stockh) 1991;71:495–501.
6 Peters MS, Winkelmann RK, Greaves MWE, Kephart GM, Gleich GJ: Extra cellular deposition of eosinophil granule major basic protein in pressure urticaria. J Am Acad Dermatol 1987;16:513–517.
7 Kontou-Fili K, Maniatakou G, Demaka P, Gonianakis M, Palailogos Y, Aroni K: The therapeutic effects of cetirizine in delayed pressure urticaria: Clinicopathologic findings. J Am Acad Dermatol 1991;24:1090–1093.
8 Shelley WB, Rawnsley HM: Aquagenic urticaria: Contact sensitivity reaction to water. JAMA 1964;189:895–898.
9 Shelley WB, Shelley ED: Adrenergic urticaria: A new form of stress-induced hives. Lancet 1985;ii:1031–1035.
10 Juhlin L, Malmros-Enander I: Solar urticaria: Mechanism and treatment. Photodermatology 1986;3:164–168.
11 Duschet P, Schwarz T, Gschnait F: Plasmapherese bei Lichturtikaria. Ein rationales Therapiekonzept in Fällen mit nachgewiesenem Serumfaktor. Hautarzt 1989;40:553–555.
12 Grattan CEH, Hamon CGB, Cowan MA, Leeming RJ: Preliminary identification of a low molecular weight serological mediator in chronic idiopathic urticaria. Br J Dermatol 1988;119:179–184.
13 Henocq E, Vargraftig BB: Accumulation of eosinophils in response to intracutaneous PAF-acether and allergens in man. Lancet 1986;i:1373–1379.
14 Hammarslund A, Pipkorn U, Enerbäck L: Mast cells tissue histamines and eosinophils in early- and late-phase skin reactions: Effects of a single dose of prednisolone. Int Arch Allergy Appl Immunol 1990;93:171–177.
15 Frew AJ, Kay AB: Eosinophils and T-lymphocytes in late phase allergic reactions. J Allergy Clin Immunol 1990;85:533–538.
16 Juhlin L, Pihl-Lundin I: Effects of antihistamines on cutaneous reactions and influx of eosinophils after local injection of PAF, kallikrein, compound 48/80 and histamine in patients with chronic urticaria and healthy subjects. Acta Derm Venereol (Stockh) 1992;72:197–200.
17 Peters MS, Schroeter AL, Kephart GM, Gleich GJ: Localisation of eosinophil granule major basic protein in chronic urticaria. J Invest Dermatol 1983;81:39–43.
18 Leiferman KM, Norris PG, Murphy GM, Hawk JLM, Winkelmann RK: Evidence for eosinophil degranulation with deposition of granule major basic protein in solar urticaria. J Am Acad Dermatol 1989;21:75–80.
19 Lessof MH, Gant V, Hinuma K, et al: Recurrent urticaria and reduced diamine oxidase activity. Clin Exp Allergy 1990;20:373–376.
20 Shelley WB, Shelley DE: Advanced Dermatological Therapy. Philadelphia, Saunders, 1987, pp 502–513.
21 Juhlin L: Recurrent urticaria: Clinical investigation of 330 patients. Br J Dermatol 1981;104:369–381.
22 Kennard CD, Ellis CN: Pharmacologic therapy for urticaria. J Am Acad Dermatol 1991;25:176–189.
23 Juhlin L, Landor M: Drug therapy of chronic urticaria. Clin Rev Allergy 1992;10:1–21.
24 Fradin MS, Ellis CN, Goldfarb MT, Voorhees JJ: Oral cyclosporine for severe chronic idiopathic urticaria and angioedema. J Am Acad Dermatol 1991;25:1065–1067.

Lennart Juhlin, MD, PhD, Department of Dermatology, University Hospital,
S–751 85 Uppsala (Sweden)

H_1-Antihistamines as Broad-Spectrum Drugs for the Treatment of Various Allergic Disorders

Jean-Pierre Rihoux

UCB s.a., Braine-l'Alleud, Belgium

The allergic reaction appears today as a very complicated phenomenon involving a large number of interacting cells, i.e. T cells, neutrophils, eosinophils, monocytes, macrophages, Langerhans cells, fibroblasts, epithelial cells, endothelial cells, neurons, etc., as well as numerous mediators such as histamine, PAF, prostaglandins and thromboxanes, leukotrienes, neuromediators, and cytokines.

In front of such a complicated network of cells and such a cocktail of interacting mediators, it becomes difficult to hope that an antagonist limited to only one particular mediator can be considered as a useful pharmacological approach for therapeutic purposes in severe allergic conditions.

In the early 1980s, two drugs characterized by significant H_1-blocking properties, ketotifen and oxatomide, were shown to inhibit the histamine release phenomenon from basophils and mast cells [1–3]. The presence in these molecules of such DSCG-like activities was soon considered as a probable advantage for the treatment of more severe allergic disorders, and more especially for the treatment of asthma.

Since that time, other drugs characterized by H_1-blocking properties (azelastine, terfenadine, loratadine) were also tested in the same way and were also shown to display similar antidegranulating effects in vitro, and again the presence of such properties in these molecules was interpreted in terms of improved therapeutic profile [4–9].

The discovery of antidegranulating properties in a series of H_1-antihistamines prompted the searchers to extensively investigate this family of chemicals in order to evidence other pharmacological activities that could also be considered as advances as far as the therapeutic index is considered. In table 1, we try to

Table 1. Listing of the main tests used for the pharmacological investigation of second-generation H_1-blockers

Inhibition of histamine release from mast cells and basophils in vitro	ketotifen oxatomide azelastine terfenadine loratadine
Inhibition of LTC_4 release from various cells in vitro [7–11]	azatadine terfenadine loratadine
LTC_4, LTD_4 antagonism [12, 13]	azelastine ketotifen
PAF antagonism [12, 13]	azelastine ketotifen
In vitro inhibition of IgG-induced eosinophil activation [14]	ketotifen
In vitro inhibition of fMLP, PAF, C5a, LTB_4 and IL-8-induced eosinophil chemotaxis [15; Sehmi et al., in press]	cetirizine
In vitro inhibition of ASA and antigen-induced platelet cytotoxicity [16]	cetirizine
In vitro inhibition of calmoduline [17]	azelastine
In vitro inhibition of calcium mobilization [8, 18–20, 28]	oxatomide ketotifen azelastine loratadine terfenadine
In vitro potentiation of β_2 response [21]	cetirizine
In vitro inhibition of PAF-induced adherence of eosinophils to endothelial cells [Sehmi et al., in press]	cetirizine
In vitro inhibition of INF_γ-induced ICAM-1 expression on epithelial cells [22]	cetirizine
In vivo inhibition of antigen-induced eosinophils' accumulation in atopics [23]	cetirizine
In vivo inhibition of antigen-induced late histamine release in atopics [24]	cetirizine
In vivo inhibition of antigen-induced mediators (histamine or kinins or PGD_2 or LTC_4) release in nasal fluid of atopics [10, 25–27]	azatadine terfenadine loratadine cetirizine

Table 2. Aspects of the investigation to be taken into consideration

Concentrations	Is the drug tested at physiological concentrations? If not, what is the ratio in comparison with drug plasma levels?
Drug	Is the drug metabolized, and if yes, is (are) the main metabolite(s) also tested?
Material	Is the test performed in vitro, ex vivo or in vivo? Is it performed on human material? Is it performed on human allergic material?

summarize a series of these additional activities that have been described with the second-generation H_1-antihistamines.

The impressive listing in table 1 of additional pharmacological activities described with these drugs might suggest that the second-generation H_1-antihistamines can really be considered as broad-spectrum drugs for the treatment of severe allergic disorders such as asthma or atopic dermatitis. It also suggests that these compounds are different from each other in expressing these unexpected activities and hence should be characterized by different clinical efficacy. In other words, it is the pharmacological profile of the drug which makes us believe that one drug could be better adapted for the treatment of an allergic condition than the other one.

While carefully examining the various methodologies used in the pharmacological (or clinical pharmacological) investigation of this family of drugs (as expressed in table 1), we focused our attention on a series of aspects which probably deserve a special mention and which are summarized in table 2. These considerations led us to propose the following three-step approach in order to try and establish some value scale to be used for an easier comparison of the drugs often presented as therapeutic advances on the basis of their pharmacological profiles.

First step: Are the pharmacological data *relevant?* Relevant means: Described in vivo in allergic subjects (or ex vivo on human material) at an acceptable dose of the drug.

Second step: Are the pharmacological activities *pertinent?* Pertinent means: Predictive of an antiallergic activity in humans.

Third step: If one drug displays several pharmacological activities, is the combination of activities *coherent?* Coherent means: Are all these activities pertinent, and better are there any potentiating activities in the pharmacological profile? Does the combination especially correspond to a well-defined allergic condition?

If it is quite easy to make a statement about the first step, the second and third steps, in contrast, remain highly speculative, and there still remains much to do before we can give reliable answers about the 2 last points. In the meantime, only the clinical data are of value and must be considered by the clinician as safe guides for the choice of a treatment in front of serious allergic disorders.

References

1 Levander S, Hägermark Ö, Stähle M: Peripheral antihistamine and central sedative effects of three H_1-receptor antagonists. Eur J Clin Pharmacol 1985;28:523–529.
2 Church MK, Gradidge CF: Oxatomide: Inhibition and stimulation of histamine release from human lung and leucocytes in vitro. Agents Actions 1980;10:4–7.
3 Radermecker M: Inhibition of allergen-mediated histamine release from human cells by ketotifen and oxatoide. Respiration 1981;41:45–55.
4 Chand N, Pillar J, Diamantis W, Sofia RD: Inhibition of IgE-mediated allergic histamine release from rat peritoneal mast cells by azelastine and selected antiallergic. Agents Actions 1985;16:318–322.
5 Little MM, Casale TB: Azelastine inhibits IgE-mediated human basophil histamine release. J Allergy Clin Immunol 1989;83:862–865.
6 Chand N, Pillar J, Diamantis W, Sofia RD: Inhibition of allergic histamine release by azelastine and selected antiallergic drugs from rabbit leukocytes. Int Arch Allergy Appl Immunol 1985;77:451–455.
7 Nabe M, Agrawal DK, Sarmiento EU, Townley RG: Inhibitory effect of terfenadine on mediator release from human blood basophils and eosinophils. Clin Exp Allergy 1989;19:515–520.
8 Akagi M, Mio M, Miyoshi K, Tasaka K: Antiallergic effects of terfenadine on immediate type hypersensitivity reactions. Immunopharmacol Immunotoxicol 1987;9:257–279.
9 Kreutner W, Chapman RW, Gulbelkian A, Siegel MI: Antiallergic activity of loratadine, a non-sedating antihistamine. Allergy 1987;42:57–63.
10 Togias AG, Naclerio RM, Warner J, Proud D, Kagey-Sobotka A, Nimmagadda I, Norman PS, Lichtenstein LM: Demonstration of inhibition of mediator release from human mast cells by azatadine base. Jama 1986;255:225–229.
11 Chand N, Pillar J, Nolan K, Diamantis W, Sofia RD: Inhibition of allergic and nonallergic leukotriene C4 formation and histamine secretion by azelastine: Implication for its mechanism of action. Int Arch Allergy Appl Immunol 1989;90:67–70.
12 Katayama S, Tsunoda H, Sakuma Y, Kai H, Tanaka I, Katayama K: Effect of azelastine on the release and action of leukotriene C_4 and D_4. Int Arch Allergy Appl Immunol 1987;83:284–289.
13 Achterrath-Tuckermann U, Weischer Ch, Szelenyi I: Azelastine, a new antiallergic/antiasthmatic agent, inhibits PAF-acether-induced platelet aggregation, paw edema and bronchoconstriction. Pharmacology 1988;36:265–271.
14 Kishimoto T, Sato T, Ono T, Takahashi K, Kimura I: Effect of ketotifen on the reactivity of eosinophils with the incubation of anti-IgG. Br J Clin Pharmacol 1990;44:226–230.
15 Leprevost C, Capron M, De Vos Ch, Tomassini M, Capron A: Inhibition of eosinophil chemotaxis by a new antiallergic compound (cetirizine). Int Arch Allergy Appl Immunol 1988;87:9–13.
16 De Vos C, Joseph M, Leprevost C, Vorng H, Tomassini M, Capron M, Capron A: Inhibition of human eosinophil chemotaxis and of the IgE-dependent stimulation of human blood platelets by cetirizine. Int Arch Allergy Appl Immunol 1989;88:212–215.
17 Middleton E, Ferriola P, Drzewiecki G, Duane Sofia R: The effect of azelastine and some other antiasthmatic and antiallergic drugs on calmodulin and protein kinase C. Agents Actions 1989;28:9–15.
18 Tasaka K, Akagi M, Mio M, Miyoshi K, Nakaya N: Inhibitor effects of oxatomide on intracellular Ca mobilization, Ca uptake and histamine release, using rat peritoneal mast cells. Int Arch Allergy Appl Immunol 1987;83:348–353.

19 Mio M, Izashi K, Tasaka K: Substance P-induced histamine release from rat peritoneal mast cells and its inhibition by antiallergic agents and calmodulin inhibitors. Immunopharmacology 1991;22:59–66.
20 Nakamura T, Nishizawa Y, Sato T, Yamato C: Effect of azelastine on the intracellular Ca^{2+} mobilization in guinea pig peritoneal macrophages. Eur J Pharmacol 1988;148:35–41.
21 Advenier C, Candenas M-L, Naline E, De Vos C: The effect of cetirizine on the human isolated bronchus: Interaction with salbutamol. J Allergy Clin Immunol 1991;88:104–113.
22 Canonica GW, Parodi MN, Pesce GP, Bagnasco M: Adhesion molecules on epithelial cells. Allergy Clin Immunol News 1991(suppl 1, A 152):128.
23 Fadel R, Herpin-Richard N, Rihoux J-P, Henocq E: Inhibitory effect of cetirizine 2 HCl on eosinophil migration in vivo. Clin Allergy 1987;17:373–379.
24 Michel L, De Vos C, Dubertret L: Cetirizine effects on the cutaneous allergic reaction in humans. Ann Allergy 1990;65:512–516.
25 Bousquet J, Lebel B, Chanal I, Morel A, Michel F-B: Antiallergic activity of H_1-receptor antagonists assessed by nasal challenge. J Allergy Clin Immunol 1988;82:881–887.
26 Naclerio RM, Kagey-Sobotka A, Lichtenstein LM, Freidhoff L, Proud D: Terfenadine, an H_1-antihistamine, inhibits histamine release in vivo in the human[1–3]. Am Rev Respir Dis 1990;142:167–171.
27 Naclerio RM, Proud D, Kagey-Sobotka A, Freidhoff L, Norman PS, Lichtenstein LM: The effect of cetirizine on early allergic response. Laryngoscope 1989;99:596–599.
28 Tasaka K, Mio M, Okamoto M: Intracellular calcium release induced by histamine releasers and its inhibition by some antiallergic drugs. Ann Allergy 1986;56:464–469.

Dr. J.-P. Rihoux, UCB s.a., Medical Department/S3, Chemin du Foriest,
B–1420 Braine-l'Alleud (Belgium)

Specific Immunotherapy

Jean Bousquet, François-Bernard Michel

Clinique des Maladies Respiratoires, Centre Hospitalier Universitaire, Montpellier, France

Immunotherapy (SIT) was introduced in 1911 and remains in many countries one of the most prescribed treatment of adolescents and young adults. SIT was found ineffective in the 1970s but significant progress has been made in the past 15 years, especially for the treatment of allergies to *Hymenoptera* venoms [1] and pneumoallergens [2–4]. However, using potent extracts, allergen injections resulted in a greater number of systemic reactions and possibly of deaths [5, 6]. Thus, SIT is again contested and threatened with some decline for pneumoallergens. On the other hand, SIT cannot be replaced by any other treatment in venom allergy.

Mechanisms

The mechanisms of action of SIT are still unclear and several possibilities have been proposed. For decades it has been postulated that allergen-specific IgG or IgG_4 present in serum and secretions may block allergens before their interaction with cell bound IgE. This blocking activity may have some relevance in the protective effect of venom immunotherapy (VIT) but is unlikely to be a major mechanism with inhalant allergens. The possible decrease in specific IgE cannot explain either the benefits of SIT.

A decrease in the end-organ reactivity has been observed using allergen-specific skin tests, conjunctival, nasal and bronchial challenges [2–4]. Both the early and the late phase reactions are inhibited. These changes appear very rapidly after the onset of SIT and far before any specific IgG change. The release of inflammatory mediators in body fluids is decreased after treatment [7] and these changes were correlated with the efficacy of SIT [8]. Finally, it has been elegantly demonstrated that eosinophils were down-activated [9] during SIT possibly because SIT decreases the generation of chemotactic factors.

These changes in cell sensitivity may be due to cytokines since it has been recently shown that after 3 days of VIT, a serum factor decreased the sensitivity of platelets to venom. During pollen SIT mononuclear cells synthetize less histamine-releasing factors. Unpublished studies have examined the effect of SIT on the Th1-Th2 phenotype of T cells and have shown that skin tests with allergen elicited a late-phase reaction in which IL-4 and IL-5 were released whereas in a proportion of subjects after SIT, the late reaction was decreased and this was accompanied by the presence of IFN-γ suggesting a switch from Th_2 to Th_1 type cells.

Immunotherapy in Venom Allergy

Efficacy

Venom immunotherapy (VIT) with standardized extracts was introduced in the 1970s and two controlled prospective studies clearly have shown that venoms are highly effective in the prevention of systemic allergic reactions. Guidelines for treatment were indicated in the late seventies and it appeared that a monthly dose of 100 μg of venom should protect most patients [1]. A multicentric study in the USA including more than 1,500 patients treated with venom according to these guidelines showed than over 85% of them are totally protected when stung by living insects. However, a small proportion is not completely protected and treatment failures have been reported [10].

Although most patients can reach the maintenance dose, some highly sensitive subjects with abnormal immune responses or alternatively patients with increased skin mast cells cannot reach this dose. During the initial phase of VIT a few patients present recurrent SR making it impossible to reach the maintenance dose. The venom doses administered are usually too low to elicit an IgG immune response. It was therefore proposed to combine active and passive immunotherapy by administering passively hyperimmune IgG fractionated from a beekeeper's serum pool to patients having a low IgG response to improve the safety of VIT in some patients [1].

Safety

Venoms are potent standardized allergen extracts and systemic reactions (SR) are not rare during VIT. Rush protocols may increase the rate of SR. Vespid venom IT appears to be better tolerated than the honey bee venom one [4]. β-Blocking agents are increasing the risks of VIT [11].

Duration

The duration of VIT is another matter of debate. It has been shown that a significant proportion of patients lose their sensitivity within 1–3 years (negative skin tests and no detectable serum IgE). In these patietns, VIT can be safely stopped. It is moreover suggested that after of course of VIT of 5 years, patients may stop VIT [12].

Immunotherapy in Pollen-Rhino-Conjunctivitis

Objectives

The treatment of allergic diseases combines immunologic and pharmacologic treatments. In many patients, drugs can relieve nasal and ocular symptoms without side effects. In these patients, SIT is not indicated. However, in the subset of patients in whom drugs are not sufficiently effective or induce side effects SIT may be started.

However, differences between the pharmacologic and immunologic treatments of allergy should not be restricted to safety and efficacy. Drugs only represent a symptomatic treatment whereas SIT might cure the disease if it is shown that the effects of SIT last after its cessation. Finally, SIT may prevent the onset of asthma in patients suffering from nasal symptoms since it was observed in may studies that nasal symptoms often precede bronchial ones.

Efficacy of Subcutaneous Immunotherapy

The efficacy of subcutaneous SIT has been demonstrated by double-blind placebo-controlled studies with standardized or polymerized extracts in rhinitis [2–4] and conjunctivitis [13] caused by grass or ragweed pollen. For most other pollen species data are lacking and the precise value of SIT is not yet fully determined. Although SIT was shown to be effective in optimal conditions, this treatment completely fails when inappropriately used and SIT is not equally effective in all patients.

The efficacy of SIT is dose-dependent. Low doses are usually ineffective and should not be used any longer. On the other hand, the 'highest tolerated dose' increases the risk of severe systemic reactions and should be replaced by the optimal maintenance dose, i.e. the dose inducing a change in cell sensitivity but giving a low and acceptable rate of mild systemic reactions [4].

It was shown that patients only allergic to grass pollens differ clinically and immunlogically from those allergic to many pollen species [14] and it was shown in a double-blind placebo-controlled study that SIT was more effective in patients allergic to a single pollen species only than in those allergic to many pollen species.

Safety of Subcutaneous Immunotherapy

Life-threatening reactions are not rare with high quality extracts and deaths have been exceptionally reported. The rate of SR is greater with standardized pollen extracts than with either nonstandardized extracts or high-molecular-weight preparations but does not differ from SIT performed with standardized extracts of other allergen species [4]. The occurrence of systemic reactions was studied in over 500 grass pollen allergic individuals who received the same rush SIT protocol with the same standardized extracts [15]. The rate of bronchial SR was increased in patients who had presented asthma during the pollen season but not in those who did not. Polymerized high-molecular-weight preparations were shown consistently to induce few systemic reactions in grass, ragweed and tree pollen allergy.

Duration

Duration of SIT is another matter of debate but it appears that a long-term treatment, at least 3 years, is required. In a retrospective study, Mosbech and Østerballe [16] observed that the effect of grass pollen SIT lasted several years after its cessation. Grammer et al. [17] using high-molecular-weight polymerized extracts reached similar conclusions. However, prospective controlled studies are lacking and no definite conclusion can be made.

Availability of Suitable Extracts

International Reference Preparations (IRP) and International Standards (IS) have been prepared under the auspices of the Allergen Subcommittee of IUIS/WHO. They include many pollens, e.g. timothy and Bermuda grass, short ragweed, birch and mugwort. Standardized extracts of many grass, weed and tree pollens are available in many parts of the world. These extracts are of known potency and although they may be more difficult to administer than low potency extracts, they should be preferred. Standardized extracts adsorbed on alum are available in Europe. Polymerized extracts have been used for trials but, except in some European countries, they are not available.

Other Routes for Allergen Administration

The usual route of administration of SIT is the subcutaneous one. Recent studies have, however, shown that oral or nasal SIT is effective when the appropriate extracts are given at a high dose. Systemic reactions were rather low with local administration of extracts but focal reactions are not rare, especially using the nasal route. Oral SIT with current extracts is not cost-effective since very large doses of extracts are required. The addition of oral SIT to booster parenteral SIT may be effective but more data are needed. Thus, these two routes need further investigations to be proposed routinely in clinical practice [4]. Sublingual SIT is currently under investigation but definite results are not available yet.

Immunotherapy in Asthma

Objectives

Asthma is a complex disease in which allergic and nonallergic triggers are interacting and result in bronchial obstruction and inflammation. The role of inhalant allergens is asthma exacerbations has been clearly demonstrated both for perennial allergens such as house dust mites or insect dusts and seasonal allergens such as grass pollens or molds as well as for animal danders. The inhalation of allergens leads to a complex activation of various cell types and the release of inflammatory and neurogenic mediators and cytokines. In allergic asthma, two different situations seem to exist. House dust mites and other perennial allergens induce a long-sustained inflammation of the bronchi leading to a variable degree of nonspecific bronchial hyperreactivity (BHR). After a long course of the disease, inflammation is a major cause of symptoms and BHR, suggesting that the immunologic treatment should be replaced by anti-inflammatory drugs. On the other hand, bronchial inflammation and BHR are usually transient in patients who are only allergic to pollens [2].

The objective of immunologic treatment is in the short term ro reduce the allergic triggers precipitating symptoms, and in the long term to decrease bronchial inflammation and BHR when it is not too severe [18].

Pollen Asthma

Controlled studies have shown that with an appropriate extract and schedule SIT in grass pollen asthma is effective [2, 3]. Aqueous and standardized extracts or formaldehyde allergoids were shown to be effective on asthma during the pollen season. Moreover, bronchial symptoms were better relieved than nasal ones. On the other hand, fewer studies found that SIT induced no significant protection. Three studies examining the effect of SIT on bronchial challenges observed that the threshold dose eliciting an immediate positive challenge was significantly increased after SIT and the late phase reaction was ablated. SIT decreases bronchial reactivity in birch pollen asthmatics. These observations indicate clearly that grass pollen asthma is improved by SIT, but only when optimal conditions are achieved. These conclusions are, however, now restricted to grass, birch and ragweed pollen allergies.

House Dust Mite Asthma

Efficacy. Although some studies are conclusive, house dust extracts should not be used any longer. Using aqueous or standardized *Dermatophagoides pteronyssinus* and/or *Dermatophagoides farinae* extracts, many studies found a significant effect of SIT [2, 3]. However, the results are not always impressive, and, especially in adults, they are sometimes negative. With other extracts, results are more variable. Bousquet et al. [19] examined who are the candidates for mite SIT. Two hundred and fifteen patients were enrolled in a controlled trial and followed up for 1 year by means of symptom-medication scores and pulmonary function. Patients who had other perennial allergies, or aspirin intolerance and/or chronic sinusitis did not improve. Moreover, among the patients allergic only to *D. pteronyssinus,* children presented a significantly greater improvement than adults, and patients with irreversible airflow limitation (FEV_1 under 70% from predicted values after an adequate pharmacologic treatment) did not benefit from SIT. Mosbech et al. [20] using either a standardized extract or a polyethylene glycol-modified extract did not reach similar conclusions but many patients were treated with inhaled corticosteroids during the whole trial and the interpretation of the data was therefore more difficult.

Machiels et al. [21] using allergen-antibody complexes made from allergens of *D. pteronyssinus* and an excess of purified autologous specific antibodies claimed that this treatment was safe and highly effective as assessed by symptoms scores and bronchial challenge with allergen. However, the complexes are difficult to be prepared and until human monoclonal antibodies are available it will be difficult to apply this technique to a large scale.

Safety. The safety of mite SIT is critical since many if not most patients who recently died from an allergen injection were asthmatics and death was caused by an irreversible bronchial obstruction. Although some investigators suggest that SIT is indicated in more severe patients, Bousquet's group performed a controlled study in over 1,000 patients who received the same maintenance dose using a rush or a step procedure. Most patients with FEV_1 under 70% of predicted values developed asthma during SIT. The age of the patient also appears to be of importance. Children under 5 years of age present a significantly greater risk of systemic reaction [22].

Duration. The duration of SIT with mite extracts is again a subject of debate. Price et al. [23] stopped SIT after a year of treatment and noticed that most children who stopped SIT presented a relapse within a year. Using SIT with standardized extracts, Knani's group [24] studied the duration of efficacy of SIT after its cessation in perfectly controlled asthmatics. The rate of relapse was dependent on the duration of SIT. When SIT was performed between 18 and 35 months, the rate of relapse was 65 vs. 35% in patients with a SIT duration of over 36 months.

Availability of Suitable Extracts. D. pteronyssinus IS are available. In many countries standardized extracts from *D. pteronyssinus* and *D. farinae* are available. These extracts are usually in lyophilized or aqueous form but, in Europe, they are often adsorbed on alum.

Allergy to Animal Proteins

It has been shown that SIT with animal dander may be effective [3, 25] using challenge and few clinical studies.

Allergy to Molds

Molds are major allergens in asthma but they often induce polysensitizations and most extracts are not standardized yet. SIT using standardized extracts of *Alternaria* [26] and *Cladosporium* [27] was shown to be effective. However, SIT with many mold species or with extracts of unknown quality should not be administered.

Immunotherapy with Other Extracts

Although SIT may be administered by oral or sub-lingual routes, there is no controlled study showing that it is effective in asthma. Specific immunotherapy with extracts of undefined allergens (bacteria, foods, *Candida albicans,* inset dusts) should not be used any longer [18, 28, 29].

Indications of Immunotherapy Depending on Recent Guidelines

Venom Allergy

Venom immunotherapy is essential for the protection of patients who have presented a severe anaphylactic reaction and in whom an IgE-mediated allergic reaction has been demonstrated by skin tests and/or serum-specific IgE. In children and possibly in adults with a mild systemic reaction in the form of generalized urticaria, VIT may be discussed [1].

Pollen Allergy

Clinical efficacy does not mean clinical indication, especially since pharmacologic treatment is also available for the treatment of allergic diseases. The specific indications of pollen SIT have been extensively studied and the position papers of the EAACI [28] and WHO/IUIS [29] have proposed some guidelines. It is commonly accepted that SIT is indicated in severe pollinosis. It also appears that SIT is indicated when asthma complicates rhinoconjunctivitis. On the other hand, SIT should not be started in mild-to-moderately severe rhinoconjunctivitis responding favorably to antihistamines and topical drugs, except if the season is long-lasting as may be the case in Southern Europe, South Africa or California. Varney et al. [13] pointed out that patients with perennial asthma should be specifically excluded, but this recommendation appears to be important only if asthma is moderately severe or severe.

Perennial Asthma

According to the Guidelines of the International Consensus on the Diagnosis and Management of Asthma some recommendations for the use of SIT in asthma have recently been proposed [18]. They are not very different from those of

Table 1. Considerations for initiating immunotherapy [adapted from Bush et al., 31]

1. Presence of a demonstrated IgE-mediated disease
   Positive skin tests and/or serum-specific IgE
2. Documentation that specific sensitivity is involved in asthma symptoms
   Exposure to the allergen(s) determined by allergy testing
     related to appearance of syptoms
   If required bronchial challenge with the relevant allergen(s)
3. Characterization of other triggers that may be involved in asthma symptoms
4. Severity and duration of asthma symptoms
   Subjective symptoms
   Objective parameters, e.g. work loss, school absenteeism
   Pulmonary function (essential): *exclude* patients with severe asthma
   Monitoring of the pulmonary function by peak flow
5. Response of asthma symptoms to nonimmunologic treatment
   Response to allergen avoidance
   Response to pharmacotherapy
6. Availability of standardized or high-quality extracts
7. Age of the patient
   Children over 5
   Adults under 50
8. Contraindications
   Treatment with β-blocker
   Other immunologic disease
   Inability of patients to comply
9. Sociologic factors
   Cost
   Occupation of candidate
   Asthma impairing quality-of-life despite adequate pharmacologic treatment
10. Objective evidence of efficacy of immunotherapy for the selected patient (availability of controlled clinical studies)

reports from WHO/IUIS [29] and EAACI [28]. In all these recommendations the quality of the extract to be administered was stressed. Optimally, these extracts should be standardized using the most modern techniques [30].

The indication of SIT should be carefully proposed in highly selected patients (table 1). It is therefore essential to carefully select the patients and consider several factors in order to appreciate the respective value of SIT in comparison with other available therapeutic methods including allergen avoidance and pharmaco-

Table 2. Recommendations to minimize risk and improve efficacy of SIT in asthma [from International Consensus Report on Diagnosis and Management of Asthma, 18]

1	Specific immunotherapy needs to be prescribed by specialists and adminstered by physicians trained to manage systemic reactions if anaphylaxis occurs
2	Patients with multiple sensitivities and/or nonallergic triggers may not benefit from specific immunotherapy
3	Specific immunotherapy is more effective in children and young adults than later in life
4	It is essential for safety reasons that patients should be asymptomatic at the time of the injections because lethal adverse reactions are more often found in asthma patients with severe airways obstruction
5	FEV_1 with pharmacologic treatment should reach at least 70% of the predicted values, for both efficacy and safety reasons

therapy: (1) Potential severity of the affection to be treated. (2) Efficacy of available treatments. (3) Optimal use of SIT using standardized extracts, high-quality modified extracts or purified components of known composition and efficacy. (4) Cost and duration of each type of treatment. (5) Risk incurred by the patient due to the allergic disease and the treatment. (6) Modification of the course of the disease by each type of treatment.

To minimize risk and improve efficacy some suggestions have been made (table 2). Absolute contraindications include patients with other serious immunopathological conditions, malignancies, poor compliance, severe psychological disorders and/or treatment by β-blocking agents, even when administered topically.

Before initiating SIT, avoidance of exposure to the allergen(s) causing asthma should always be attempted. Except in the case of animal dander, most common aeroallergens cannot be avoided completely and this is particularly true for patients allergic to house dust mites and those who are allergic to multiple allergens.

Allergy to house dust mites may be influenced by mite avoidance measures that should always be applied. If such measures fail and if pharmacotherapy is incompletely effective or leads to side effects, SIT may be envisaged but only if the mites are the only perennial allergen. The indications of SIT in mite allergy have been extensively examined but a definite conclusion cannot be made until the results of the large study performed by the American Academy of Allergy and Immunology and NIAID are available. At present it is suggested that SIT is effective in children and young adults with a strict mite allergy and in whom FEV_1 after pharmacologic treatment is over 70% of predicted values.

In animal dander allergy, allergen avoidance is the best choice, but SIT to cat or dog may occasionally be an alternative in occupational allergy and in some children to whom the eviction of the animal may be an intense shock.

In mold asthma, the elimination of indoor allergen is favored and SIT may be restricted to patients only allergic to *Alternaria* and /or *Cladosporium*.

Conclusions

Specific immunotherapy was introduced in 1911 and for many years remained based on empiricism. Although this form of treatment has been seriously criticised, it still represents an effective treatment of allergic diseases. Administered with care by specialists it has gained an increased safety. Using high-quality extracts, SIT may be envisaged in severe pollinosis and mite asthma, and its use may be questioned in animal dander and mold allergy. New forms of SIT are made available but oral and sublingual routes deserve further studies. Speculations on the prevention of asthma by administering SIT in patients with rhinitis are currently being tested. New technologies might lead to improved forms of immunologic treatment and immunotherapy may be one of the treatments of the future.

References

1   Bousquet J, Müller UR, Dreborg S, Jarish R, Malling HJ, Mosbech H, et al: Immunotherapy with *Hymenoptera* venoms. Allergy 1987;42:707–720.
2   Van Metre TE Jr, Adkinson NF Jr: Immunotherapy for aeroallergen disease; in Middleton E Jr, Reed CE, Ellis EF, Adkinson NF Jr, Yunginger JW (eds): Allergy, Principles and Practice, ed 3. St Louis, Mosby, 1989, pp 1327–1344.
3   Bousquet J, Hejjaoui A, Michel FB: Specific immunotherapy in asthma. J Allergy Clin Immunol 1990;86:292–306.
4   Bousquet J, Michel FB: Specific immunotherapy in rhinitis, clinical aspects; in Mygind N, Naclerio RN (eds): Vasomotor and Allergy Rhinitis, ed 2. Copenhagen, Munksgaard, 1992, in press.
5   Committee on Safety of Medicine. Desensitizing vaccines. Br Med J 1986;293:948.
6   Norman PS: Fatal misadventures. J Allergy Clin Immunol 1987;79:572–573.
7   Ilioupoulos O, Proud D, Adkinson NF, et al: Effects of immunotherapy on the early, late and rechallenge nasal reaction to provocation with allergen: Changes in inflammatory mediators and cells. J Allergy Clin Immunol 1991;87:855–866.
8   Bousquet J, Maasch HJ, Martinot B, et al: Double-blind placebo controlled immunotherapy with mixed grass pollen allergoids. II. Comparison between parameters assessing the efficacy of immunotherapy. J Allergy Clin Immunol 1988;82:439–446.
9   Rak S, Löwhagen O, Venge P: The effect of immunotherapy on bronchial hyperresponsiveness and eosinophil cationic protein in pollen-allergic patients. J Allergy Clin Immunol 1988;82:470–480.
10 Bousquet J, Ménardo JL, Velasquez G, Michel FB: Systemic reactions during maintenance immunotherapy with honey bee venom. Ann Allergy 1988;61:63–68.
11 Awai LE, Mekri YA: Insect sting anaphylaxis and β-adrenergic blockade: A relative contra-indication. Ann Allergy 1984;53:48–49.

12 Bousquet J, Knani J, Velasquez G, Ménardo JL, Guilloux L, Michel FB: Evolution of sensitivity to hymenoptera venoms in 200 allergic patients followed for up to 3 years. J Allergy Clin Immunol 1989;83/84:944–950.

13 Varney VA, Gaga M, Aber VR, Kay AB, Durham SR: Usefulness of immunotherapy in patients with severe summer hay fever uncontrolled by antiallergic drugs. Br Med J 1991;302:265–269.

14 Bousquet J, Becker WM, Hejjaoui A, Cour P, Chanal I, Lebel B, Dhivert H, Michel FB: Clinical and immunological reactivity of patients allergic to grass pollens and to multiple pollen species. II. Efficacy of a double-blind, placebo-controlled, specific immunotherapy with standardized extracts. J Allergy Clin Immunol 1991;88:43–53.

15 Hejjaoui A, Ferrando R, Michel FB, Bousquet J: Systemic reactions occurring during immunotherapy with standardized pollen extracts. J Allergy Clin Allergy, in press.

16 Mosbech H, Østerballe O: Does the effect of immunotherapy last after termination of treatment? Allergy 1988;43:523–529.

17 Grammer LC, Shaughnessy MA, Suszko JM, et al: Persistence of efficacy after a brief course of polymerized ragweed allergens: A controlled study. J Allergy Clin Immunol 1984;73:484–489.

18 International Consensus Report on Diagnosis and Management of Asthma. Allergy 1992;47 (suppl 1).

19 Bousquet J, Hejjaoui A, Clauzel AM, et al: Specific immunotherapy with a standardized *Dermatophagoides pteronyssinus* extract. II. Prediction of efficacy of immunotherapy. J Allergy Clin Immunol 1988;79:971–977.

20 Mosbech H, Dreborg S, Frølund J, et al: Hyposensitization in asthmatics with mPEG modified and unmodified house dust mite extract. II. Effect evaluated by challenges with allergen and histamine. Allergy 1989;44:499–509.

21 Machiels JJ, Somville MA, Lebrun PM, Lebecque SJ, Jaquemin MG, Saint-Rémy JMR: Allergic bronchial asthma due to *Dermatophagoides pteronyssinus* can be efficiently treated by inoculation of allergen-antibody complexes. J Clin Invest 1990;85:1024–1035.

22 Hejjaoui A, Dhivert H, Michel FB, Bousquet J: Specific immunotherapy with a standardized *Dermatophagoides pteronyssinus* extract. IV. Systemic reactions according to the immunotherapy schedule. J Allergy Clin Immunol 1990;85:490–497.

23 Price JF, Warner JOP, Iley EN, Turner MW, Soothill JF: A controlled trial of hyposensitization with adsorbed tyrosine *Dermatophagoides pteronyssinus* antigen in childhood asthma: In vivo aspects. Clin Allergy 1984;14:209–220.

24 Hejjaoui A, Knani J, Dhivert H, Michel FB, Bousquet J: Duration of efficacy of specific immunotherapy with a standardized mite extract after its cessation. J Allergy Clin Immunol 1992;89:319.

25 Van Metre TE, Marsh DG, Adkinson NF Jr: Immunotherapy for cat asthma. J Allergy Clin Immunol 1988;82:1055–1068.

26 Horst M, Hejjaouui A, Horst V, Michel FB, Bousquet J: Double-blind, placebo-controlled rush immunotherapy with a standardized *Alternaria* extract. J Allergy Clin Immunol 1990;85:460–472.

27 Malling HJ, Dreborg S, Weeke B: Diagnosis and immunotherapy of mould allergy. V. Clinical efficacy and side effects of immunotherapy with *Cladosporium herbarum*. Allergy 1986;41:507–519.

28 Malling HJ, et al: Position paper of the European Academy of Allergy and Clinical Immunology: Specific immunotherapy. Allergy 1988(suppl 6):431–463.

29 Thompson R, Bousquet J, Cohen S, Frei PC, Jager L, Lambert PH, Lessof MH, Loblay RH, Malling HJ, Norman PS, De Weck AL, Weeke B: The current status of allergen immunotherapy (hyposensitization). Report of WHO/IUIS Working Group. Lancet 1989;i:259–261.

30 Reed CE, Yunginger JW: Quality assurance and standardization of allergy extracts in allergy practice. J Allergy Clin Immunol 1989;84:4–8.

31 Bush RK, Huftel MA, Busse WW: Patient selection; in Lockey RF, Bukantz SC (eds): Allergen Immunotherapy. New York, Marcel Dekker, 1991, pp 25–49.

Jean Bousquet, MD, PhD, Clinique des Maladies Respiratoires, Centre Hospitalier Universitaire, F-34000 Montpellier (France)

Inhibitors of Leukotriene Synthesis and Actions: Asthma Drugs of the Near Future?

Rodger M. McMillan

Vascular Inflammatory and Musculo Skeletal Research Department, ZENECA Pharmaceuticals, Mereside, Alderley Park, Macclesfield, Cheshire, UK

Introduction

Metabolism of arachidonic acid (AA) by 5-lipoxygenase leads to the formation of a group of biologically active lipids known as leukotrienes. Peptido-leukotrienes (peptido-LTs) have spasmogenic activity in airways smooth muscle and also contribute to allergic inflammation by enhancing vascular permeability. Leukotriene B_4 (LTB_4) is a potent chemotactic agent for a variety of leukocytes and induces leukocyte-dependent oedema. In view of these properties, leukotrienes have been proposed as important mediators in allergic and inflammatory disorders and leukotriene antagonists or inhibitors of 5-lipoxygenase may have therapeutic potential in a range of diseases including asthma. This review will describe the 5-lipoxygenase pathway and the biological properties of leukotrienes which implicate them as potential mediators in asthma. Progress in the discovery and development of peptido-LT antagonists and 5-lipoxygenase inhibitors will be discussed. Finally, the efficacy of leukotriene antagonists and 5-lipoxygenase inhibitors in asthmatics will be summarised.

5-Lipoxygenase and Leukotrienes

Lipoxygenase pathways are widely distributed in the plant and animal kingdoms and are responsible for catalysing the oxidative metabolism of a range of unsaturated fatty acids. Lipoxygenases have been extensively studied and characterised. Indeed, soyabean lipoxygenases was one of the first enzymes to be prepared in pure crystalline form. In mammalian systems the predominant substrate is arachidonic acid (AA) and lipoxygenase are classified according to the positional specificity for oxidation of AA. The blood platelet 12-lipoxygenase, the first reported mammalian lipoxygenase, was first described in 1974 [1] but its physiological significance is still unclear. In contrast, the mammalian 5-lipoxygenase pathway, first described in 1979 [2], produces potent biological mediators and this has stimulated the search for 5-lipoxygenase inhibitors.

AA metabolism by 5-lipoxygenase leads to the formation of a reactive hydroperoxide, 5-hydroperoxy-eicosatetraenoic acid (5-HPETE). Reduction of 5-HPETE can occur non-enzymatically or may be catalysed by peroxidases and results in formation of the corresponding hydroxy acid, 5-hydroxy-eicosatetraenoic acid (5-HETE). Alternatively, 5-HPETE can itself serve as a substrate for 5-lipoxygenase and is converted to an unstable epoxide intermediate, leukotriene A_4 (LTA_4), which has a pivotal position and may be converted to either a 5, 12-di-hydroxy acid, leukotriene B_4 (LTB_4), or to peptido-LTs (LTC_4, LTD_4 or LTE4). The term leukotriene was introduced to denote the fact that the molecules are produced in *leuko*cytes and all contain conjugated *trienes*.

Leukotriene synthesis has been demonstrated in all classes of leukocytes. However, the profile of 5-lipoxygenase products depends on the species, source and leukocyte type. Thus, polymorphonuclear leukocytes synthesise predominantly LTB_4 whilst peptido-LTs are the major products in mast cells. Murine peritoneal macrophages synthesize LTC_4 but little LTB_4 whereas rabbit alveolar macrophages produce relatively large amounts of LTB_4. There is approximately 93% sequence homology at the amino acid level between human and rat 5-lipoxygenase [2, 3]. Moreover, there is considerable homology with soyabean and mammalian 15-lipoxygenases. In particular, there are 5 conserved histidine residues, 2 of which are presumed, from site-directed mutagenesis studies, to contribute to the iron-binding domain of the active site [4].

5-Lipoxygenase has been purified from several sources. Molecular masses of approximately 78 kD have been determined [2, 3] and in each case activity is dependent on calcium and ATP, a feature which distinguishes the enzymes from other lipoxygenases [5]. Despite the calcium dependence of 5-lipoxygenase no sequence homologies have been found with calcium-binding domains in calmodulin and other calcium-binding proteins. In resting cells, 5-lipoxygenase is cytosolic and kinetic studies have employed this soluble form of the enzyme. On activation of leukocytes, 5-lipoxygenase undergoes a calcium-dependent translocation and associates with an 18-kD membrane protein, 5-lipoxygenase-activating protein (FLAP) [6]. Transfection experiments have conclusively demonstrated that expression of both 5-lipoxygenase and FLAP is required for cellular leukotriene biosynthesis [7]. It is presumed that this 'docking' phenomenon localises

5-lipoxygenase close to the source of substrate arachidonic acid, which is generated from membrane phospholipids. However, the precise mechanism of interaction of 5-lipoxygenase and FLAP remains to be determined.

Elevated levels of leukotrienes are detected in the blood of asthmatics during acute attacks [8, 9] and leukotrienes are present in bronchoalveolar lavage fluid from asthmatic patients [10]. Recently, there has been considerable interest in measuring levels of leukotrienes in urine as a non-invasive measure of changes in 5-lipoxygenase activity. Elevated concentrations of LTE_4 have been demonstrated in a number of situations including patients with acute exacerbations of asthma, allergen challenge of asthmatics and aspirin challenge of aspirin-sensitive asthmatics [11, 12].

Peptido-Leukotrienes: Mediators of Airways Obstruction and Inflammation

Peptido-LTs have powerful spasmogenic actions particularly in airway smooth muscle and in the vasculature. In general their contractile responses can be distinguished from those of other agonists in 2 ways. First, they are remarkably potent being at least 100 times more potent on a molar basis than histamine [13]. Second, responses to peptido-LTs are slower in onset and of more prolonged duration than those of other spasmogens and LTC_4, LTD_4 and LTE_4 collectively account for all the biological actions previously ascribed to 'slow-reacting substance of anaphylaxis' (SRS-A) [14]. Although guinea pig tracheal and parenchymal smooth muscle have been widely used for studying peptido-LTs, these differ in 2 significant ways from the human airways [15]. In guinea pigs, a substantial component of the contractile response to peptido-LT is mediated indirectly through stimulation of release of thromboxane A_2 but this does not occur in humans. Also, there are at least 2 distinct peptido-LT receptors in guinea pig airways; one mediates responses to LTC_4 and the other is an LTD_4/LTE_4 receptor. In contrast, human bronchi possess a single peptido-LT receptor which mediates responses to LTC_4, LTD_4 and LTE_4 [15]. Also, in human airways, the 3 peptido-LTs are virtually equipotent but LTD_4 is usually more potent than LTC_4 and LTE_4 in the guinea pig.

In accord with their in vitro potencies described in the previous section, peptido-LTs are more than 100 times more potent than histamine or methacholine as bronchoconstrictors in man when administered by aerosol. Moreover, asthmatic patients show enhanced sensitivity to the bronchoconstrictor effects of peptido-LT; thus normal volunteers are 40 times more sensitive to LTD_4 than platelet-activating factor (PAF) but in asthmatics LTD_4 is 1,000 times more potent than PAF [15].

Peptido-LTs may also influence airway obstruction by stimulating mucus secretion. Evidence for enhanced mucus secretion was obtained following in vivo administration of LTC_4, LTD_4 or LTE_4, as well as other lipoxygenase products [16, 17]. This is supported by in vitro studies using human airways in which peptide leukotrienes increase mucus production, with LTD_4 being the most potent stimulant in the human model [18].

Another property of peptide leukotrienes, with relevance to inflammatory disorders such as asthma, is enhanced vascular permeability. LTC_4, LTD_4 and LTE_4 have been shown to induce plasma protein exudation from the microvasculature in a range of species [19–21]. In humans, intradermal injection of peptide leukotrienes induces classical wheal and flare reactions [22]. The relevance of these inflammatory actions to the lung was recently demonstrated when LTD_4 was shown to induce leakage of plasma proteins in the guinea pig respiratory tract [23].

Leukotriene B$_4$: A Potent Mediator of Leukotaxis and Inflammation

LTB_4 does not directly contract airways smooth muscle although it induces an indirect response in guinea pig parenchyma which is mediated through the release of thromboxane A_2. However, LTB_4 is a potent mediator of inflammation due to its chemotactic activity in neutrophils, eosinophils, monocytes and lymphocytes. Most studies have been performed with neutrophils in which its chemotactic potency is comparable to peptide chemotactic factors such as FMLP and C5a. In common with other chemotactic agents, LTB_4 stimulates chemokinesis, chemotaxis, aggregation and degranulation [24, 25]. LTB_4 also stimulates adherence of neutrophils to endothelium both in vitro and in vivo [26]. The early observations that geometric and stereochemical isomers of LTB_4 as well as LTB_4 metabolites were less potent activators of PMN demonstrated that these effects were mediated via specific LTB_4 receptors. Two classes of LTB_4 receptors have been identified on human neutrophils [27]. High-affinity receptors are activated by subnanomolar concentrations and these mediate calcium mobilisation, chemokinesis and chemotaxis. The lower affinity receptors mediate degranulation and oxidative metabolism. There are species differences in LTB_4 responsiveness. Thus, LTB_4 does not induce chemotaxis in rat neutrophils and this has been shown to be due to the lack of the high-affinity class of receptors [28].

LTB_4 induces inflammatory reactions in animals and man. These effects are consistent with the in vitro leukotactic properties of the lipid. LTB_4 has been reported to induce PMN accumulation into human skin [29, 30] and into a variety of tissues in experimental animals including lung [31, 32]. Most studies have focussed on the contribution of LTB_4 to neutrophil-dependent inflammation. In rabbit skin it was demonstrated that LTB_4, in common with other chemotactic agents, synergises with vasodilators to produce leakage of plasma proteins [33]. Administration of arachidonic acid to rabbit skin produces a neutrophil-dependent oedema which is mediated largely by the interaction of LTB_4 and prostanoid vasodilators [34]. LTB_4 induces other signs of PMN activation in vivo including transient neutropaenia following intravenous administration and adherence and diapedesis of PMN in hamster cheek pouch [32, 35].

Of particular relevance to asthma is the ability of LTB_4 to induce eosinophil chemotaxis. In view of the considerable mediator redundancy in chemotaxis, it might be argued that inhibition of a single mediator would not be effective at inhibiting cell accumulation in a complex disease such as asthma. This remains to be determined in man but it has been demonstrated that a selective antagonist of LTB_4 (U-75,302) can inhibit allergen-induced eosinophil recruitment in the guinea pig [36].

Development of Peptide Leukotriene Antagonists

A variety of receptor antagonists of peptido-leukotrienes have been developed. The first such compound FPL55712 (Fisons) was reported as an antagonist of SRS-A before the discovery of leukotrienes [37]. FPL55712 was a key compound in characterising the in vitro biology of peptide leukotrienes but it is relatively weak and its short half-life limits its utility in vivo. Following the characterisation of SRS-A as a mixture of leukotrienes, a number of companies developed hydroxy acetophenones based on the structure of FPL55712. These represented a first generation of leukotriene antagonists and some analogues showed 10-fold improvements in potency over the parent molecule (e.g. LY171883, Lilly; L648051, Merck Frosst) [38, 39] but none had sufficient efficacy in human volunteers to provide clinical utility. A 'second generation' of very potent antagonists has now been developed. These are structurally distinct from the hydroxy acetophenones and they can be categorised into 3 chemical classes.

Structural Analogues of LTD_4/LTE_4

SKF104353 (SmithKline Beecham) and LY170680 (Lilly) are potent and selective peptido-leukotriene antagonists in vitro [45, 56]. They are approximately 100-fold more potent than FPL55712 and do not have significant effects on other agonists such as histamine. Both compounds have acceptable in vivo activity as leukotriene antagonists when administered by aerosol but they have poor oral activity. Following aerosol administration to human subjects SKF104353 and LY170680 produced modest inhibition of LTD_4-induced bronchoconstriction [47, 48].

Indole-Based Heterocyclic Amines

Accolate (ICI 204219; ZENECA) and its structural analogue, ICI198615, are potent, competitive antagonists at peptido leukotriene receptors [44, 45]. They are more than 1,000 times more potent than FPL55712 as LTD_4 antagonists and do not exhibit any inhibitory effects at other receptors at concentrations up to 10,000 times higher than those required to block leukotriene receptors. Accolate has potent in vivo activity against LTD_4-induced bronchoconstriction when administered orally, intravenously or by aerosol. It also inhibits the leukotriene component of allergic bronchospasm in guinea pigs [44]; in this model animals are pre-treated with pyrillamine, propranolol and indomethacin prior to antigen administration. The encouraging pre-clinical profile of the compound was confirmed when it was evaluated against LTD_4-induced bronchoconstriction in human volunteers. A single oral dose of 40 mg produced more than a 100-fold shift in the LTD_4 dose-response curve and significant antagonism was maintained for at least 24 h [46].

Quinoline-Based Antagonists

MK571 (previously referred to as L660711) has high affinity for peptido-LT receptors and is approximately 1000 times more potent than FPL55712 [47]. The compound inhibits LTD_4-induced bronchoconstriction when administered by oral, intravenous or aerosol routes. MK571 also inhibits antigen-induced dyspnea in rats and allergic bronchospasm in *Ascaris*-sensitive squirrel monkeys. Intravenous infusion of MK571 in human asthmatics, at plasma concentrations of 2 and 20 µg/ml, produced, respectively, at least 40- and 80-fold shifts in the dose-response curve for LTD_4-induced bronchoconstriction [48].

Development of 5-Lipoxygenase Inhibitors

The therapeutic potential of 5-lipoxygenase inhibitors in asthma as well as other inflammatory diseases has stimulated intense activity in the pharmaceutical industry. The search for 5-lipoxygenase inhibitors has proved to be a much more difficult task than originally anticipated. This has resulted partly from the unusual characteristics of the enzyme. Kinetic studies are complicated by features such as apparent substrate inhibition, wich may be related to the limited aqueous solubility of substrate, and product-mediated enzyme inactivation. In addition, leukotriene synthesis in intact cells involves membrane translocation of 5-lipoxygenase and therefore conventional cell-free assay, which employ cytoplasmic enzyme preparations, can yield misleading data on the potency and selectivity of inhibitors.

In vivo evaluation of lipoxygenase inhibitors has also been problematic. Misleading data were generated from early studies because inhibitors were evaluated in animal models without first confirming their biochemical efficacy in vivo. Thus, the anti-inflammatory activity of benoxaprofen was attibuted to lipoxygenase inhibition but subsequent studies revealed that, although benoxaprofen inhibited leukotriene synthesis in some in vitro systems, it did not directly inhibit 5-lipoxygenase in vitro [49] and did not inhibit leukotriene synthesis in vivo [50]. In most cases in vivo efficacy of lipoxygenase inhibitors is now initially assessed biochemically by measuring drug effects on leukotriene synthesis ex vivo. The simplest system for these studies involves stimulation of whole blood with a calcium ionophore which induces synthesis of leukotrienes and cyclo-oxygenase products such as prostaglandin E_2 and thromboxane B_2. This approach allows measurement of potency and selectivity of inhibitors both in vitro and ex vivo and has been applied to a range of species including humans. The whole blood assay is generally used to provide the first clinical information on biochemical efficacy in volunteer studies.

At least 50 companies have patents claiming lipoxygenase inhibitors. They can be classified into 4 main categories.

(1) Redox Inhibitors

5-Lipoxygenase is presumed to be an iron-containing enzyme which is rendered inactive by reduction of Fe(III) to Fe(II). The majority of inhibitors have the potential to inhibit the enzyme by redox mechanisms. In general, these compounds have either lacked adequate selectivity or oral activity. The prototype inhibitor in this area is the dual inhibitor of cyclo-oxygenase and lipoxygenase, BW755C, which inhibits leukotriene synthesis both in vitro and in vivo and was a useful pharmacological tool [51]. However, studies with the compound in animal models were complicated by its inhibition of prostaglandin synthesis and by potential effects on other oxidative enzymes.

Redox inhibitors with selectivity for 5-lipoxygenase have been reported as illustrated by our experience with substituted indazolinones including ICI 207,968 [52]. In a variety of in vitro systems ICI 207,968 is at least 200 times more potent as an inhibitor of synthesis of leukotrienes than of PGE_2. This high degree of selectivity is combined with oral potency which permits meaningful studies on the biological roles of leukotrienes without concerns about the contribution of prostaglandins. ICI 207,968 produces dose-dependent inhibition of the peptido-LT component of antigen-induced bronchoconstriction in sensitised guinea pigs. The compound also inhibits leukocyte accumulation and plasma exudation induced by AA in the rabbit, a model previously shown to be mediated in part by LTB_4. Activity of indazolinones in this model correlates with lipoxygenase inhibitor potency [52]. The disadvantages of compounds of this type are also illustrated with ICI 207,968. Thus, ICI 207,968 has cytoprotective properties in the gastro-intestinal tract [53] but it is not clear whether this is related to lipoxygenase inhibition or is a consequence of the redox properties of the compound. Interference with other oxidative enzymes may also have severe toxicological consequences and a common problem with many redox inhibitors, including ICI207968, is induction of methaemoglobinaemia. The author is not aware of any redox inhibitors currently undergoing clinical evaluation.

(2) Hydroxamates and N-Hydroxy Ureas

The discovery that hydroxamic acid analogues of AA were lipoxygenase inhibitors stimulated activity on related compounds and numerous patents have appeared. Wellcome published a series of papers on acetohydroxamates including BWA4C. These compounds are characterised by modest in vitro selectivity (10- to 30-fold) for 5-lipoxygenase compared to cyclo-oxygenase and good oral activity [54]. The blood levels achieved after oral dosing are insufficient to inhibit cyclo-oxygenase and therefore BWA4C causes selective inhibition of leukotriene synthesis ex vivo with reasonable duration. Acetohydroxamates have oral activity against the peptido-LT component of allergic bronchospasm in the guinea pig [55]. In addition, these compounds inhibit leukocyte accumulation in rat inflammatory exudate but this effect does not correlate strongly with lipoxygenase inhibition [56]. BWA4C inhibited ex vivo LTB_4 synthesis when dosed orally to volunteers [57] but further development was terminated due to extensive metabolism and unacceptable accumulation of metabolites.

A related series of compounds are the N-hydroxy ureas from Abbott exemplified by zileuton (A-64077). Zileuton has similar in vitro potency and selectivity to the acetohydroxamates and inhibits leukotriene synthesis ex vivo [63]. In the dog this effect is prolonged due to reduced metabolism. Zileuton inhibits allergic bronchospasm and has anti-inflammatory activity in rodents, including inhibition of leukocyte accumulation [58].

Zileuton produces dose-dependent inhibition of leukotriene synthesis ex vivo following oral administration to volunteers. The duration of action is relatively short with doses of 600-800 mg zileuton inhibiting leukotriene synthesis for only 6 h [59]. Thus, single doses produce intermittent inhibition and maintenance of inhibition requires dosing q.i.d. with 600 mg zileuton. Recently, Abbott reported another N-hydroxy urea (A78773) which exhibits a longer half-life in experimental animals and in man [60]. The compound has been administered to volunteers and inhibition of leukotriene synthesis for 24 h has been achieved with once daily dosing [61].

(3) Non Redox Inhibitors of 5-Lipoxygenase

The compounds described above have the potential to participate in redox reactions or to ligand to iron. Since the mechanism of 5-lipoxygenase is believed to involve an iron-catalysed redox cycle it is likely that such properties contribute to the mechanism of action of acetohydroxamates and N-hydroxy ureas. A theoretical concern with these agents is that they may have actions on a range of other iron-containing enzymes. Although there are no reports of toxicological problems with zileuton as a consequence of these effects, it is of interest that evidence for a specific interaction of N-hydroxy ureas and acetohydroxamates with 5-lipoxygenase is lacking. In particular, there is no difference in enzyme inhibitor potency with enantiomers of optically active agents from these structural classes.

A novel series of 5-lipoxygenase inhibitors, methoxyalkyl-thiazoles, were developed at ICI/Zeneca. These compounds do not have the potential to participate in redox reactions or to chelate iron and represent a unique class of non-redox inhibitors. In contrast to redox inhibitors and iron-liganding inhibitors, 5-lipoxygenase inhibition by methoxyalkyl-thiazoles exhibits tight structure activity relationships [62]. Moreover, using optically active analogues, there is evidence for enantioselective inhibition of 5-lipoxygenase [68]. Methoxyalkyl-thiazoles, such as ICI211965 and ICI216800, are useful research tools but they lacked sufficient oral bioavaibility to permit clinical evaluation.

Structural modification of the methoxyalkyl-thiazole series led to the development of analogues with improved potency and oral activity. These include methoxy-tetrahydropyrans, exemplified by ICI D2138, which is the most potent and most selective inhibitor of 5-lipoxygenase inhibitor yet published [64]. ICI D2138 produces prolonged inhibition of leukotriene synthesis ex vivo and inhibits allergic bronchospasm and AA-induced inflammation [64]. ICI D2138 inhibits ex vivo leukotriene synthesis when administered orally to volunteers [65]. The half-life of the compound in man (ca. 12 h) is sufficient to permit 24-hour suppression of leukotriene synthesis with once daily administration.

(4) Inihibitors of Translocation

A series of compounds with a novel mode of action were developed by Merck Frosst. These agents, exemplified by MK886 (L663536), do not inhibit 5-lipoxygenase but prevent translocation of the enzyme to the membrane. This appears to be mediated by antagonism of the interaction of 5-lipoxygenase with 5-lipoxygenase activating protein (FLAP) [6,7] MK886 is a highly selective compound with no effects on prostaglandin synthesis in vitro or in vivo. The compound is remarkably potent against leukotriene generation in plasma-free leukocyte suspensions but potency is reduced 100-fold in whole blood indicating extensive plasma protein binding. MK886 is orally active as a leukotriene synthesis inhibitor with excellent duration of action and the compound inhibits antigen-induced bronchoconstriction in *Ascaris*-sensitive squirrel monkeys [66].

Following the elucidation of the mechanism of action of MK886, other agents that inhibit leukotriene synthesis have been shown to be FLAP antagonists, including a Revlon compound, REV5901 (now PF5901; Purdue Frederick) and WY50295 (Wyeth Ayerst). Both agents produce selective inhibition of leukotriene synthesis and WY50295 is orally active as an inhibitor of leukotriene synthesis in experimental animals [67]. However, both compounds failed to inhibit leukotriene synthesis in volunteers. Recently, Merck Frosst reported a series of molecules which are hybrids between the previous quinoline and indole analogues [68]. These 'quindoles', such as MK591, are also FLAP antagonists and inhibit leukotriene synthesis in vitro and ex vivo. MK591 inhibits allergic bronchospasm in rat, guinea pig and monkey. Volunteer data with MK591 demonstrated potent, prolonged inhibition of ex vivo leukotriene synthesis and the pharmacodynamics of the compound are compatible with one daily dosing for 24-hour suppression of leukotriene synthesis [69].

Evaluation of Leukotriene Modulators in Asthmatics

The foregoing sections reviewed the progress in development of peptido-LT antagonists and inhibitors of leukotriene biosynthesis. Examples of both classes of drug have now reached clinical evaluation in asthmatics. Most clinical data are available with the leukotriene antagonists, notably MK571 and Accolate.

Intravenous administration of MK571 produced attenuation of exercise-induced bronchoconstriction in patients with stable asthma [70]. MK571 also inhibited both early and late responses to inhaled antigen [71]. Acute improvements in lung function have also been reported following administration of MK571 to asthmatics [72]. In a 6-week study of mild-to-moderate asthma, an enantiomer of MK571, MK679 (500 mg t.i.d.), significantly improved FEV_1 (forced expiratory volume in 1 s) as well as daytime and nocturnal symptoms [73].

The promising pre-clinical profile of Accolate has been confirmed in asthmatic patients. Oral administration produced inhibition of early and late responses to allergen [74, 75] and of exercise-induced bronchoconstriction [76]. In a 14-day in-patient study, Accolate (60 or 100 mg once daily) produced significant improvement in FEV_1 [77]. In a subsequent out-patient study, 6 weeks' treatment with 10 mg b.i.d. produced significant clinical benefit [77].

Zileuton is the only inhibitor of leukotriene synthesis for which clinical data in asthma have been published. A single oral dose of 800 mg zileuton produced significant improvement of cold air-induced bronchoconstriction in asthmatics

[78]. This dose also reduced nasal congestion induced by nasal challenge with allergen, a model of allergic rhinitis [79]. However, in another study, zileuton was not effective in inhibiting bronchoconstriction induced by inhaled allergen [80]. A subsequent 28-day study in asthma, using zileuton at 600 mg q.i.d., resulted in significant improvements in FEV_1 and symptom score as well as a reduction in bronchodilator usage [81].

Conclusions

There is now compelling evidence that modulation of leukotrienes will provide a successful therapeutic approach in bronchial asthma. Leukotrienes can contribute to lung function by direct effects on airways smooth muscle, modulation of mucus secretion and influencing vascular and cellular aspects of inflammation. Elevated leukotriene levels are detected in blood, lavage fluid and urine of active asthmatics. There has been considerable progress in the development of agents that modulate the synthesis or action of leukotrienes. Encouraging efficacy in asthma has been reported with 2 peptido-LT antagonists and one of these (Accolate) is undergoing further clinical evaluation. A 5-lipoxygenase inhibitor, zileuton, has also demonstrated efficacy in asthma and its development is proceeding. Examples of 3 classes of leukotriene synthesis inhibitor (ICI D2138, A78773 and MK591) are also in clinical development. All provide improved potency and duration of action compared to zileuton and offer the potential for once-daily dosing. An assessment of the available data leads to the conclusion that the hypothesis for leukotriene involvement in asthma is proved. Drugs that modify the synthesis or actions of leukotrienes are poised to provide the first novel therapeutic advance in asthma treatment for over 20 years.

References

1. Nugteren DH: Inhibition of prostaglandin biosynthesis by 8 cis, 12 trans, 14 cis-eicosatrienoic acid and 5 cis, 8 cis, 12 trans, 14 cis-eicosatetraenoic acid. Biochim Biophys Acta 1975;380:290–307.
2. Dixon RAF, et al; Cloning of the complementary DNA for human 5-lipoxygenase. Proc Natl Acad Sci USA 1988;85:416–420.
3. Balcarek JM, et al: Isolation and characterisation of a cDNA clone encoding rat 5-lipoxygenase. J Biol Chem 1988;263:13937–13941.
4. Percival MD, Ouellet M: The characterization of 5 histidine-serine mutants of human 5-lipoxygenase. Biochim Biophys Res Commun 1990;173:507–513.
5. Rouzer CA., Samuelsson B: On the nature of the 5-lipoxygenase reaction in human leukocytes: Enzyme purification and requirement for multiple stimulatory factors. Proc Natl Acad Sci USA 1985;82:6040–6044.
6. Ford-Hutchinson AW: FLAP: A novel drug target for inhibiting the synthesis of leukotrienes. Trends Pharmacol Sci 1991;12:68–70.

7   Dixon RAF, et al: Requirement of a 5-lipoxygenase-activating protein for leukotriene synthesis. Nature 1990;343:282–284.
8   Isono T, Koshihara Y, Murota S, Fukuda Y, Furukaw S: Measurements of immunoreactive leukotriene C_4 in blood of asthmatic children. Biochem Biophys Res Commun 1985;130:486–492.
9   Zakrzewski JT, Barnes NC, Piper PJ, Costello JF: Measurement of leukotrienes in arterial and venous blood from normal and asthmatic subjects by radioimmunoassay. Br J Clin Pharmacol 1985;19:574P.
10   Wardlaw AJ, et al: Leukotrienes LTC_4 and LTB_4 in bronchial asthma and other respiratory diseases. J Allergy Clin Immunol 1989;84:19–26.
11   Kumlin M, et al: Urinary excretion of leukotriene E4 and 11-dehydro-Thromboxane B_2 in response to brochial provocations with allergen, aspirin, leukotriene D_4, and histamine in asthmatics. Am Rev Resp Dis 1992;146:98–103.
12   Smith CM, et al: Urinary leukotriene E_4 in bronchial asthma. Eur Respir J 1992;5:693–699.
13   Dahlen S-E, et al: Allergen challenge of lung tissue from asthmatics elicits bronchial contraction that correlates with the release of leukotrienes C_4 D_4 and E_4. Proc Natl Acad Sci USA 1983;80:1712–1716.
14   Lewis RA, et al: Slow reacting substances of anaphylaxis: Identification of leukotriene C-1 and leukotriene D from human and rat sources. Proc Natl Acad Sci USA 1980;77:3710–3714.
15   O'Donnell M, Welton A: Comparison of the pulmonary pharmacology of leukotrienes and PAF: Effects of their antagonists; in Lewis AJ, Doherty NS, Ackeman NR (eds): Therapeutic approaches to Inflammatory Diseases. New York, Elsevier, 1989, pp 169–193.
16   Johnson HG, et al: Leukotriene-C_4 enhances mucus production from submucosal glands in canine trachea in vivo. Int J Immunopharmacol 1983;5:391–396.
17   Yanni JM, et al: Effect of intravenously administered lipoxygenase metabolites on rat tracheal mucous gel layer thickness. Int Arch Allergy Appl Immunol 1989;90:307–309.
18   Marom Z, et al: Slow reacting substances, leukotrienes C_4 and D_4, increase the release of mucus from human airways in vitro. Am Rev Respir Dis 1982;126:449–451.
19   Peck MJ, Piper PJ, Williams TJ: The effects of leukotrienes C_4 and D_4 on the microvasculature of guinea pigs. Prostaglandins 1981;21:315–321.
20   Ueno A, Tanaka K, Katori M, Arai Y: Species differences in increased vascular permeability by synthetic leukotriene C_4 and D_4. Prostaglandins 1981;21:637–648.
21   Dahlen SE, et al: Leukotrienes promote plasma leakage and leukocyte adhesion in postcapillary venules. In vivo effects with relevance to the acute inflammatory response. Proc Natl Acad Sci USA 1981;78:3887–3891.
22   Camp RDR, et al: Responses of human skin to intradermal injection of leukotrienes C_4, D_4 and B_4. Br J Pharmacol 1980;80:497–502.
23   Evans TW, et al: Regional and time dependent effects of inflammatory mediators on airway microvascular permeability in the guinea pig. Clin Sci 1989;76:479–485.
24   Ford-Hutchinson AW, et al: Leukotriene B: A potent chemokinetic and aggregating substance released from polymorphonuclear leukocytes. Nature 1980;286:264–265.
25   Rollins TE, et al: Synthetic leukotriene B_4 is a potent chemotaxin but a weak secretagogue for human PMN. Prostaglandins 1983;25:281–289.
26   Lindbom L, et al: Leukotriene B_4 induces extravasation and migration of polymorphonuclear leukocytes in vivo. Acta Physiol Scand 1982;116:105–108.
27   Goldman DW, Goetzl EJ: Heterogeneity of human polymorphonuclear leukocyte receptors for leukotriene B_4. J Exp Med 1984;159:1027–1041.
28   Kreisle RA, et al: Specific binding of leukotriene B_4 to a receptor on human polymorphonuclear leukocytes. J Immunol 1985;134:3356–3363.
29   Soter NA, Lewis RA, Corey EJ, Austen KF: Local effects of synthetic leukotrienes (LTC_4, LTD_4, LTE_4 and LTB_4) in human skin. J Invest Dermatol 1983;80:115–119.
30   Camp RDR, et al: Production of intraepidermal microabscesses by topical application of leukotriene B_4. J Invest Dermatol 1985;84:427–429.
31   Staub NC, Shultz EL, Koche K, Albertine: Effect of neutrophil migration induced by leukotriene B_4 on protein permeability in sheep lung. Fed Proc 1985;44:30–35.

32   Bray MA, Ford-Hutchinson AW, Smith MJH: Leukotriene B_4: An inflammatory mediator in vivo. Prostaglandins 1981;22:213–222.
33   Wedmore CV, Williams TJ: Control of vascular permeability by polymorphonuclear leukocytes in inflammation. Nature 1981;289:646–650.
34   Aked DM, Foster SJ: The contribution of eicosanoids to the acute inflammatory reaction induced by arachidonic acid in rabbit skin. Br J Pharmacol 1987;92:545–552.
35   Lindbrom L, et al: Leukotriene B_4 induced extravasation and migration of polymorphonuclear leukocytes in vivo. Acta Physiol Scand 1982;116:105–108.
36   Richards IM, Dunn CJ, Oostveen JA, Griffin RL: Effect of the selective leukotriene B_4 antagonist U-75302 on antigen-induced bronchopulmonary eosinophilia in sensitized guinea pigs. FASEB J 1989;3:A914.
37   Augstein J, et al: Selective inhibition of slow reacting substance of anaphylaxis. Nature New Biol 1973;245:215–216.
38   Fleisch JH, et al: LY171883, 1-<2-hydroxy-3-propyl-4-<4-(1-H-tetrazol-5-yl)butoxy>phenyl>ethanone, and orally active leukotriene D_4 antagonists. J Pharmacol Exp Ther 1985;233:148–157.
39   Young RN, et al: Novel arylthio and arylsulfonyl propyloxy acetophenones: Design and synthesis of L-648051 and L-649923, potent antagonists of leukotriene D_4; in Samuelsson B, Paoletti R, Ramwell PW (eds): Advances in Prostaglandins, Thromboxane and Leukotriene Research, vol 17. New York, Raven Press, 1987, pp 544–548.
40   Hay DW, et al: A novel, potent and selective peptidoleukotriene receptor antagonist in guinea pig and human airways. J Pharmacol Exp Ther 1987;243:474–481.
41   Boot JR, et al: The pharmacological evaluation of LY170680, a novel leukotriene D_4 and E_4 antagonist in the guinea-pig. Br J Pharmacol 1989;98:259–267.
42   Wood-Baker R, Turner GA, Lucas B, Holgate ST: The effect of inhaled LY170680 on leukotriene D_4-induced bronchoconstriction in normal subjects. Am Rev Respir Dis 1989;141:A115.
43   Evans HM, Piper PJ, Costello JF: The pharmacological profile of SK&F104353-72, a potent selective inhaled antagonist of cysteinyl leukotrienes, in normal man; in Samuelsson B, Ramwell R, Paolettin R, Folco G, Granstrom E (eds): Advances in Prostaglandin Thromboxane and Leukotriene Research. New York, Raven Press, vol 21A, p 469.
44   Krell RD, et al: The preclinical pharmacology of ICI 204,219 a peptide leukotriene antagonist. Am Rev Resp Dis 1990;141:978–987.
45   Snyder DW, et al: In vitro pharmacology of ICI 198,615: A novel, potent and selective peptide leukotriene antagonist. J Pharmacol Exp Ther 1987;243:548–556.
46   Smith LJ, et al: Inhibition of leukotriene D_4-induced bronchoconstriction in normal subjects by the oral LTD_4 receptor antagonist ICI 204,219. Am Rev Respir Dis 1990;141:988–992.
47   Jones TR, Zamboni R, Belley M: Pharmacology of L-660,71 (MK-571): A novel potent and selective leukotriene D_4 receptor antagonist. Can J Physiol Pharmacol 1989;67:77.
48   Kips JC, et al: MK-571, a potent antagonist of LTD_4-induced bronchoconstriction in the human. Am Rev Respir Dis 1989;139:A63.
49   McMillan RM, Masters DJ, Sterling WW, Bernstein PR: Biosynthesis of leukotriene B_4 in human leukocytes. Demonstration of a calcium-dependent 5-lipoxygenase: in Bailey JM (ed): Prostaglandins, Leukotrienes and Lipoxins. New York, Plenum Press, 1985, pp 655–668.
50   Salmon JA, Higgs GA, Tilling L, Moncads S, Vana JR: Mode of action of benoxyprofen. Lancet 1984;i:848.
51   Higgs GA, Flower RJ, Vane JR: A new approach to anti-inflammatory drugs. Biochem Pharmacol 1979;28:1959–1961.
52   Foster SJ, Bruneau P, Walker ERH, McMillan RM: 2-Substituted indazolinones: Orally active and selective 5-lipoxygenase inhibitors with anti-inflammatory activity. Br J Pharmacol 1990;99:113–118.
53   Foster SJ, Aked DM, McCormick ME, Potts HC: Cytoprotective properties of ICI207968, a selective 5-lipoxygenase inhibitor, on the rat gastrointestinal mucosa. Br J Pharmacol 1989;96:38P.
54   Tateson JE, et al: Selective inhibition of arachidonate 5-lipoxygenase by novel acetohydroxamic acids: Biochemical assessment in vitro and ex vivo. Br J Pharmacol 1988;94:528–539.

55 Payns AN, Garland LG, Lees IW, Salmon JA: Selective inhibition of arachidonate 5-lipoxygenase by novel acetohydroxamic acids: Effects on bronchial anaphylaxis in anaesthetised guinea-pigs. Br J Pharmacol 1988;94:540–546.

56 Higgs GA, Follenfant RL, Garland LG: Selective inhibition of arachidonate 5-lipoxygenase by novel acetohydroxamic acids: Effects on acute inflammatory responses. Br J Pharmacol 1988;94: 547–551.

57 Nicholls A, Posner J: Investigation of single doses of BWA4C, a selective 5-lipoxygenase inhibitor in healthy volunteers. Br J Clin Pharmacol 1991;31:577P.

58 Carter GW, Young PR, Albert DH, Bousha J, Dyer R, Bell RL, Summers JB, Brooks DW: 5-Lipoxygenase inhibitory activity of zileuton. J Pharmacol Exp Ther 1991;256:929–937.

59 Rubin P, Dube L, Braeckman R, Swanson L, Hansen R, Albert D, Carter G: Pharmacokinetics, safety, and ability to diminish leukotriene synthesis by zileuton, and inhibitor of 5-lipoxygenase; in Acherman NR, Bonney RJ, Doherty N (eds): Progress in Inflammation Research and Therapy. 1991, pp 103–112.

60 Bell RL, et al: A-78773, a new potent, orally effective, long-acting 5-lipoxygenase inhibitor. Proc 8th Int Conf Prostaglandins and Related Compounds 1992, abstr 154.

61 Rubin P, et al: Zileuton, a 5-lipoxygenase inhibitor, efficacy in human clinical trials. Proc 8th Int Conf Prostaglandins and Related Compounds, 1992, abstr 638.

62 Bird TGC, Bruneau P, Crawley GC, Edwards MP, Foster SJ, Girodeau J-M, Kingston JF, McMillan RM: Methoxyalkyl thiazoles: A new series of potent, selective and orally active 5-lipoxygenase inhibitors displaying high enantioselectivity. J Med Chem 1991;34:2176–2186.

63 McMillan RM, Girodeau JM, Foster SJ: Selective chiral inhibitors of 5-lipoxygenase with anti-inflammatory activity. Br J Pharmacol 1990;101:501–503.

64 McMillan RM, Spruce KE, Crawley GC, Walker ERH, Foster SJ: Pre-clinical pharmacology of ICI D2138, as a potent orally-active non-redox inhibitor of 5-lipoxygenase. Br J Pharmacol 1992;107: 1042–1047.

65 Yates RA, et al: A new non-redox 5-lipoxygenase inhibitor ICI-D2138 is well tolerated and inhibits blood leukoriene synthesis in healthy volunteers. Am Rev Resp Dis 1992;145:A745.

66 Gillard JA, et al: L-663,536 (MK-886) (3-[1-(4-chlorobenzyl)-3-t-butyl-thio-5-isopropylindol-2-yl]-2,2-dimethyl propanoic acid), a novel, orally active leukotriene biosynthesis inhibitor. Can J Physiol Pharmacol 1989;67:456–464.

67 Evans JF, et al: 5-Lipoxygenase-activating protein is the target of a quinoline class of leukotriene synthesis inhibitors. Mol Pharmacol 1991;40:22–27.

68 Mancini JA, et al: 5-Lipoxygenase-activating protein is the target of a novel hybrid of two classes of leukotriene biosynthesis inhibitors. Mol Pharmacol 1992;41:267–272.

69 Chervinsky P, Friedman BX, O'Neil M, Rivard S, Tanaka W, Teahan J, Zhang J: Biochemical effects of MK-0591, a potent oral leukotriene (LT) biosynthesis inhibitor in patients with mild asthma (MA). J Allergy Clin Immunol 1993;91:225.

70 Manning PJ, et al: Inhibition of exercise-induced bronchoconstriction by MK-571: A potent leukotriene D_4-receptor antagonist. N Engl J Med 1990;323:1736–1739.

71 Rasmussen JB, Eriksson L-O, Margolskee DJ, Tagari P, Williams VC, Andersson K-E: Leukotriene D_4 receptor blockade inhibits the immediate and late bronchoconstrictor responses to inhaled antigen in patients with asthma. J Allergy Clin Immunol 1992;90:193–201.

72 Gaddy JN, Margolskee DJ, Bush RK, Williams VC, Busse WW: Bronchodilation with a potent and selective leukotriene D_4 (LTD_4) receptor antagonist (MK-571) in patients with asthma. J Allergy Clin Immunol 1990;85:197A.

73 Margolskee D, et al: The therapeutic effects of MK-0679, a selective leukotriene D_4 receptor antagonist, in patients with chronic asthma. Proc 8th Int Conf Prostaglandins and Related Compounds, 1992, abstr 639.

74 Taylor IK, O'Shaughnessy KM, Fuller RW, Dollery CT: Effect of cysteinyl-leukotriene receptor antagonist ICI204219 on allegen-induced bronchoconstriction and airway hyper-reactivity in atopic subjects. Lancet 1991;337:690–694.

75 Fundlay SR, Barden JM, Easley CB, Glass M: Effect of the oral leukotriene antagonist, ICI204,219, on antigen-induced bronchoconstriction in subjects with asthma. J Allergy Clin Immunol 1992;89: 1040–1045.

76 Finnerty JP, Wood-Baker R, Thomson H, Holgate ST: Role of leukotrienes in exercise-induced asthma. Am Rev Respir Dis 1992;145:746–749.
77 Glass : ICI 204,219: Results of initial dose ranging in subjects with asthma. Proc 8th Int Conf Prostaglandins and Related Compounds, 1992, abstr 640.
78 Israel E, et al: The effects of a 5-lipoxygenase inhibitor on asthma-induced by cold, dry air. N Engl J Med 1990;323:1740–1744.
79 Knapp HR: Reduced allergen-induced nasal congestion and leukotriene synthesis with an orally active 5-lipoxygenase inhibitor. N Engl J Med 1990;323:1745–1748.
80 Hui EP, et al: Effect of a 5-lipoxygenase inhibitor on leukotriene generation and airway responses after allergen challenge in asthmatic patients. Thorax 1991;46:184–189.
81 Israel E, Drazen J, Pearlman H, Cohn J, Rubin P: A double-blind multicenter study of zileuton. A potent 5-lipoxygenase (5-LO) inhibitor versus placebo in the treatment of spontaneous asthma in adults. J Allergy Clin Immunol 1992;89:236.

Rodger M. McMillan, MD, Bioscience 1 Department, Zeneca Pharmaceuticals, Mereside, Alderley Park, Macclesfield, Cheshire SK10 4TG (UK)

Binding Sites for Peptido-Leukotrienes in Human Lung Parenchyma

Simonetta Nicosia, Valérie Capra, Serenella Giovanazzi, G. Enrico Rovati

Laboratory of Molecular Pharmacology, Insitute of Pharmacological Sciences, University of Milan, Italy

Introduction

It is nowadays believed that leukotrienes (LTs), the major components of SRS-A (slow-reacting substance of anaphylaxis), play a role in many allergic and inflammatory diseases, such as asthma, psoriasis, adult respiratory distress syndrome, rhinitis, gout, rheumatoid arthritis, inflammatory bowel disease [1]. In some allergic diseases (rhinitis, atopic dermatitis, urticaria and conjunctivitis), the role of LTs is probably secondary with respect to histamine. On the contrary, in allergic asthma it is now accepted that peptido-LTs are among the most important mediators of the bronchospasm, while LTB_4 might participate in the inflammatory aspects of the disease.

Peptido-LTs elicit their actions through specific receptors, which have been characterized pharmacologically by means of various antagonists [2]. Moreover, specific binding sites for these compounds, with the characteristics required for receptors, have been identified in many tissues, including human airways [3–6], although none of these receptors has been cloned so far.

The number and subtypes of peptido-LT receptors in the airways are still a matter of debate. The evidence available at present, which is far from definitive, has been summarized by Gardiner [7] who suggested that in guinea pig airways, one of the most widely employed models for LT investigations, two different classes of receptors exist, one that recognizes LTD_4 and LTE_4, while the other one is specific for LTC_4. In human airways, the pharmacological characterization performed so far indicates that the receptors are different from both those present in guinea pig airways and that they might recognize all the three peptido-LTs. These conclusions are based on the finding that in guinea pig airways FPL55712, ICI 198,615 and SKF 104353 are able to antagonize the contractile action of LTD_4 but not that of LTC_4; on the contrary, in human intralobar airways, the same antagonists are effective against both peptido-LTs. LTE_4 is supposed to interact with the same receptor as the other ones.

Given the potential role of peptido-LT receptor antagonists as novel antiasthma drugs, in order to design such specific and selective antagonists it is important to understand in more detail the precise nature of such receptors in human airways, and in particular to clarify whether the three peptido-LTs interact with a single homogeneous class of receptors or with heterogeneous subtypes.

We have addressed this problem by performing binding studies with a number of agonists and antagonists in membranes from human lung parenchyma. We have used an objective mathematical approach by means of computer modelling of the binding data, utilizing the computer program LIGAND [8].

Materials and Methods

Materials

[^{3}H]-ICI 198,615 (60 Ci/mmol), [^{3}H]-LTD_4 (160 Ci/mmol), were purchased from New England Nuclear (Boston, Mass.). ICI 198,615 was kindly provided by Stuart Pharmaceuticals, Division of ICI America. LTD_4, FPL55712 and Ro 24-5913 were a gift from Hoffmann-La Roche. SKF 104353 and SKF 104373 were kindly provided by Smith Kline & French Laboratories and LY171883 by Eli Lilly & Co. Serine-borate, glycine and cysteine were from Sigma.

Membrane Preparation

Membranes from human lung parenchyma were prepared as previously described [3]. Briefly, macroscopically normal specimens of human lung parenchyma were minced and homogenized at 4 °C in 50 mM Tris-HCl buffer, pH 7.4 (1:24, w:v). The homogenate was centrifuged at 770 g for 10 min (4 °C) and the supernatant centrifuged at 27,000 g for 20 min.

Binding Assay

The standard assay was performed in polypropylene tubes in a final volume of 250 μl containing 50 mM Tris-HCl buffer, pH 7.1, 0.2% ethanol, 1 mM $CaCl_2$, 0.03–0.5 nM [^{3}H]-ICI 198,615 or [^{3}H]-LTD_4; ICI 198,615, LTD_4, FPL55712, LY171883, SKF 104353, SKF 104373 and Ro 24-5913 at the indicated concentrations, serine-borate (40 mM, final concentration), glycine (10 mM, final concentration) and cysteine (10 mM, final concentration) carried with the membranes (0.15 mg of protein per sample). The incubation was carried out at 25 °C for 30 min. After the incubation, the bound ligand was separated from free by filtration through Whatman GF/C fiber glass filters.

Table 1. Binding affinities of the sites labeled by [^{3}H]-LTD$_4$

Compound	K_{d1}, nM	%CV	K_{d2}, nM	%CV
LTD$_4$	0.13	33	4,159	78
ICI 198,615	0.027	84	337	75
Ro 24-5913	27.4	51	–	
SKF 104353	11.1	47	–	
SKF 104373	–		6,277	37
FPL55712	–		7,287	52
LY171883	–		15,978	64

Table 2. Binding affinities of the sites labeled by [^{3}H]-ICI 198,615

Compound	K_{d1}, nM	%CV	K_{d2}, nM	%CV
LTD$_4$	0.13	33	4,159	78
ICI 198,615	0.027	84	337	75
Ro 24-5913	25.4	50	–	
SKF 104353	2.1	98	–	
SKF 104373	–		13,627	50
FPL55712	–		4,685	56
LY171883	–		19,628	63

Computer Analysis

Analysis of equilibrium ligand binding data were performed by using the computer program, LIGAND [8]. Selection of the best fitting model was based on the F test for the extra sum of square principle [9]. A statistical level of significance of $p < 0.01$ was selected. Data are expressed as mean ± SE.

Results

In order to determine the equilibrium binding curves of LTD$_4$ and ICI 198,615, we performed a complete self- and cross-displacement experiment for two labeled and unlabeled compounds (four curves), at the broadest possible concentration range for each compound. Computerized analysis of these data by means of LIGAND revealed the presence of two different classes of binding sites for both [^{3}H]-ICI 198,615 and [^{3}H]-LTD$_4$. Both compounds were found to recognize the same high affinity-low capacity class of sites with $K_{d1} = 0.13 \pm 0.045$ nM for LTD$_4$ and 0.027 ± 0.023 nM for ICI 198,615 and $B_{max1} = 0.007 \pm 0.0033$ pmol/mg protein, and the same low affinity-high capacity class with $K_{d2} = 4,159 \pm 3,244$ nM for LTD$_4$ and 337 ± 252.7 nM for ICI 198,615 and $B_{max2} = 21.1 \pm 16$ pmol/mg protein.

The pharmacological characterization of these binding sites was performed by means of a series of heterologous displacement curves using both [^{3}H]-ICI 198,615 and [^{3}H]-LTD$_4$ as labeled ligands and a number of unlabeled antagonists. Table 1 shows that the specific antagonists, Ro 24-5913, SKF 104353 and its enantiomer SKF 104373 as well as FPL55712 and LY171883 were found to interact with a single class of binding sites when challenged against [^{3}H]-LTD$_4$. In the same way, in table 2 we report the K_ds of the corresponding set of antagonists, this time challenged against [^{3}H]-ICI 198,615. Once again, by computer analysis, these compounds were found to recognize a homogenous class of binding sites. Comparison of tables 1 and 2 indicates a good agreement between the affinity of the various antagonists against the two different labeled ligands.

Discussion

By computer analysis of the equilibrium ligand binding data of [^{3}H]-LTD$_4$ and [^{3}H]-ICI 198,615 in human lung membranes, we have characterized a heterogeneity of binding sites for these compounds, at variance with that reported by others [10]. In fact, we report here the identification of a high affinity-low capacity class of sites with a K_d in the picomolar range and of a low affinity-high capacity class of sites with a K_d in the micromolar range. The identification of this second class of sites depends upon spanning a large range of ligand concentrations and performing the simultaneous analysis of the four curves of the complete self- and cross-displacement experiment for two ligands by means of a computer program.

On the contrary, the other antagonists tested seemed to interact with a homogeneous class of sites, either the high affinity (Ro 24-5913 and SKF 104353) or the low affinity (FPL55712, LY171883 and SKF 104373), albeit with very different affinities (tables 1, 2). The biological importance and significance of these two classes of sites remains to be established.

In conclusion, the identification and characterization of specific binding sites for [^{3}H]-LTD$_4$ and [^{3}H]-ICI 198,615 in membranes from human lung parenchyma and the finding that different LT antagonists interact with either one of the two classes of sites, might be of relevance in the development of new, more potent and selective drugs for the treatment of asthma.

References

1 Lewis RA, Austen KF, Soberman RJ: Leukotrienes and other products of the 5-lipoxygenase pathway. N Engl J Med 1990;323:645–655.
2 Halushka PV, Mais DE, Mayeux PR, Morinelli TA: Thromboxane, prostaglandin and leukotriene receptors. Annu Rev Pharmacol Toxicol 1989;10:213–239.
3 Rovati GE, Oliva D, Sautebin L, Folco GC, Welton AF, Nicosia S: Identification of specific binding sites for leukotriene C_4 in membranes from human lung. Biochem Pharmacol 1985;34:2831–2837.
4 Civelli M, Oliva D, Mezzetti M, Nicosia S: Characteristics and distribution of specific binding sites for leukotriene C_4 in human bronchi. J Pharmacol Exp Ther 1987;242:1019–1024.
5 Rovati GE, Giovanazzi S, Mezzetti M, Nicosia S: Heterogeneity of binding sites for ^{3}H-ICI 198,615 in human lung parenchyma. Biochem Pharmacol 1992;44:1411–1415.
6 Lewis MA, Mong S, Vessella RL, Crooke ST: Identification and characterization of leukotriene D_4 receptors in adult and fetal human lung. Biochem Pharmacol 1985;34:4311–4317.
7 Gardiner PJ: Eicosanoids and airway smooth muscle. Pharmacol Ther 1989;44:1–62.
8 Munson PJ, Rodbard D: LIGAND: A versatile computerized approach for characterization of ligand binding systems. Anal Biochem 1980;107:220–239.
9 Draper NR, Smith H: Applied Regression Analysis. New York, Wiley, 1966.
10 Aharony D, Falcone RC: Binding of ^{3}H-LTD$_4$ and the peptide leukotriene antagonist ^{3}H-ICI 198,615 to receptors on human lung membranes; in Zor U, Naor Z, Danon A (eds): Leukotrienes and Prostanoids in Health and Diseases. New Trends Lipid Mediators Res. Basel, Karger, 1989, vol 3, pp 67–71.

Simonetta Nicosia, Institute of Pharmacological Sciences, Via Balzaretti 9,
I-20133 Milan (Italy)

Effects of Antidiencephalon Antibody (Ser 282) on Sleep in Primary Fibromyalgia

A Preliminary Report

L. Staner[a], Ch. Kempenaers[a], T. Appelboom[b], P. Mingard[c], J. Mendlewicz[a]

[a] Sleep Laboratory, Department of Psychiatry and
[b] Rheumatology Unit, Erasme Hospital, Free University of Brussels, Belgium;
[c] Serolab, Lausanne, Switzerland

Introduction

Fibromyalgia is a nonarticular rheumatic disorder that is characterised by generalised musculoskeletal pain, fatigue, stiffness, and a number of musculoskeletal tender points in the absence of any demonstrated rheumatic disease [1]. Sleep symptoms such as unrefreshing, nonrestorative, light or restless sleep have been reported in up to 80–100% of fibromyalgia patients [2]. Moreover, a particular sleep EEG pattern, the alpha EEG non-REM anomaly, has been described in fibromyalgia syndrome (FS) [3]. Some lines of evidence have provided support for a relationship between the immune system and sleep [4] and studies have also implicated immunological functions in FS [5]. We present here preliminary results of an 8 weeks' treatment with antidiencephalon antibody (Ser 282) compared with placebo (PL) and amitriptyline (AMI) on sleep and some clinical parameters in FS. In addition to this putative involvement of immunity and sleep in FS, the rationale for using Ser 282 was that evidence has accumulated on the use of immunotherapy for the effective treatment on both psychosomatic and organic disorders [6–8].

Materials and Methods

Thirty-six female outpatients (aged 24–56) diagnosed as having FS according to standardised criteria [1] participated in the study. They were randomly allocated to one of the three treatment regimens: Ser 282 (equine immune serum against diencephalon) vs. AMI vs.

Table 1. Sleep EEG data in fibromyalgia at baseline and after treatment

Sleep variables	Baseline			Treatment			ANOVA statistics
	placebo (n = 7)	SER 282 (n = 6)	amitrip- tyline (n = 6)	placebo (n = 7)	SER 282 (n = 6)	amitrip- tyline (n = 6)	
Total sleep time (TST), min	409 (74)	364 (77)	443 (42)	449 (64)	370 (47)	421 (65)	NS
Sleep efficiency, %	85.1 (6.5)	80.3 (12.4)	89.2 (4.5)	87.1 (7.1)	85.5 (6.3)	82.5 (6.5)	NS
Sleep onset latency, min	15 (11)	21 (17)	16 (24)	12 (5)	18 (13)	22 (20)	NS
Arousal index (n/TST)	8 (3)	8 (2)	7 (2)	6 (2)	8 (3)	10 (3)	0.029[a]
Stage shifts index, (n/TST)	39 (13)	46 (6)	44 (10)	36 (12)	48 (12)	49 (10)	NS
Stage 1, min	40 (16)	39 (18)	46 (10)	44 (19)	45 (19)	56 (9)	NS
Stage 2, min	208 (40)	180 (57)	240 (33)	245 (63)	182 (43)	237 (36)	0.059[b]
Stage 3, min	32 (8)	35 (18)	42 (9)	33 (12)	31 (13)	28 (12)	NS
Stage 4, min	41 (29)	37 (33)	28 (12)	37 (21)	50 (43)	11 (14)	0.055[a]
REM sleep, min	89 (42)	73 (30)	88 (17)	90 (33)	63 (12)	89 (29)	NS
REM latency, min	82 (40)	88 (106)	69 (19)	79 (38)	50 (22)	78 (31)	NS

Mean values and (standard deviation).
[a] Group × treatment interaction; [b] group effect.

PL. All doses were stable and unchanged across the whole trial and consist of one tablet each day of AMI (50 mg) or PL and a suppository of Ser 282 (20 mg/sup) or PL 3 times a week. Psychotropic, analgesic or anti-inflammatory drugs, including corticosteroids were not allowed during the study and, if previously administered, were stopped at least 15 days before the study began (T0). T0 clinical assessments, repeated after 4 (T4) and 8 weeks (T8; end of the study), include pain-focused measures: number of positive tender points (TP), severity of localised pain (LP), polyalgia (PA) and a global self-rating of pain with a visual analogue scale (VAS). Regarding sleep, 2 VAS were filled in at T0 and T8: global sleep quality and diurnal fatigue (DF). All-night polysomnographic recordings were made during 3 consecutive nights before inclusion (T0) and during 1 night at T8. The procedure for sleep recording and definition of sleep variables are described elsewhere [9].

Results

As complete data analyses will be reported elsewhere, this section is focused on the 19 patients (6 Ser 282, 6 AMI and 7 PL) with both available and reliable baseline and T8 recording nights. Results are displayed in tables 1 (sleep data) and 2 (symptoms). Briefly, the repeated measures of ANOVAs with treatment modalities as a group factor showed the following significant results: (1) a time effect (improvement) for all pain symptoms; (2) a group effect for stage 2, despite complete randomised allocation in treatment groups; (3) interactions (time ×

Table 2. Fibromyalgia symptoms: evolution with treatments (mean and SD)

	Treatment			ANOVA statistic		
	placebo	amitriptyline	SER 282	time effect	group effect	interaction
Pain-related symptoms						
Tender points (n: 4–18)						
Baseline	16 (2)	16 (3)	16 (2)			
Treatment week 4	14 (4)	11 (4)	15 (3)	0.001	NS	0.04
Treatment week 8	11 (5)	10 (5)	14 (5)			
Severity of localized pain (0–18)						
Baseline	10 (2)	12 (1)	11 (1)			
Treatment week 4	8 (3)	7 (5)	11 (3)	0.004	NS	0.035
Treatment week 8	6 (3)	9 (4)	10 (4)			
Severity of polyalgia (0–3)						
Baseline	1.9 (0.6)	2.5 (0.6)	2.0 (0.5)			
Treatment week 4	1.5 (0.9)	1.0 (1.3)	1.9 (0.9)	0.001	NS	0.045
Treatment week 8	1.1 (1.0)	1.7 (1.0)	1.4 (1.1)			
Global pain (VAS: 0–100)						
Baseline	65 (20)	57 (23)	67 (18)			
Treatment week 4	50 (23)	42 (27)	72 (16)	0.012	NS	NS
Treatment week 8	37 (28)	32 (31)	63 (26)			
Sleep-related symptoms						
Global sleep quality (VAS: 0–100)						
Baseline	58 (7)	30 (8)	59 (39)			
Treatment week 8	60 (22)	47 (30)	36 (14)	NS	NS	NS
Diurnal fatigue (0–3)						
Baseline	2.0 (1.0)	1.5 (0.6)	2.7 (1.5)			
Treatment week 8	2.7 (1.5)	1.2 (0.5)	2.0 (1.0)	NS	NS	0.05

group) for arousal index (AI), stage 4 (ST4), TP, LP, PA and DF. AI raised with AMI, decreased with PL and was stable with Ser 282. ST4 tended to increase with Ser 282, to fall with AMI and to be unchanged with PL. DF was enhanced with PL, stable with AMI, and improved with Ser 282. For TP, LP and PA, the best improvement was achieved by AMI at T4. However, this analgesic effect was not sustained for LP and PA, leaving a moderate net gain at T8 that was lower than those obtained with PL or Ser 282. Compared to AMI, PL and Ser 282 had more gradual analgesic effects that were more obvious for PL than for Ser 282, especially for LP and TP.

Discussion

According to previous reports [10, 11], minimal symptom relief was observed with AMI in our fibromyalgia patients. Indeed, its analgesic properties at T8 were not better than those of PL or Ser 282. Thus, the three treatment conditions seemed to be responsible for the significant time effect evidence, leaving doubts about the specificity of their therapeutic potencies. The strong PL response in our patients, that could have obscured these results was already noted [10, 12] and could relate to sampling heterogeneity, spontaneous evolution as well as supportive attitude provided by our research team. The latter could have been particularly helpful in this often unrecognised or misdiagnosed [1] painful condition. T0 sleep values have already been discussed elsewhere [13]. In conformity with a recent study [14], sleep in FS was also poorly affected by treatments. Nevertheless, AMI seemed to be unexpectedly 'alerting' increasing AI and Ser 282 promoted ST4 and decreased DF. As ST4 has been implicated in body restitution [15], it can be speculated that the improvement of DF could be attributed to a ST4 enhancement effect that relates to an immune mechanism. In fact, preliminary data [16] suggest immunomodulating properties of immunosera that could be responsible of the ST4 increment described for Ser 282 in this study, via, for example, cytokines production [4].

Acknowledgements

The author wishes to thank the members of the FST (Fondation pour la Recherche et l'Application Thérapeutique des Sérums Tissulaires) for their constructive criticism concerning the design of this study.

References

1 Yunnus MB, Masi AT, Calabro JJ, Miller KA, Feigenbaum SL: Primary fibromyalgia (fibrositis): Clinical study of 50 patients with matched controls. Semin Arthritis Rheum 1981;11:151–171.
2 Campbell SM, Clark S, Tindall EA, Forehand ME, Benett RM: Clinical characteristics of fibrositis. I. A 'blinded' controlled study of tender points. Arthritis Rheum 1983;26:817–824.
3 Modolfsky H, Scarisbrick P, England R, Smythe H: Musculoskeletal symptoms and non-REM sleep disturbances in patients with 'fibrositis syndrome' and healthy subjects. Psychosom Med 1975;37:341–351.
4 Krueger JM, Toth LA, Cady AB, Johanssen L, Obal F Jr: Immunomodulation and sleep; in Inoue S, Schneider-Helmert D (eds): Sleep Peptides. Basic and Clinical Approaches. Tokyo, Japan Scientific Societies Press, 1988, pp 95–129.
5 Wallace DJ, Peter JB, Bowman RL, Wormsley SB, Silverman S: Fibromyalgia, cytokines, fatigue syndrome and immune regulation; in Fricton JR, Awad EA (eds): Myofascial Pain and Fibromyalgia. Adv Pain Res Ther. New York, Raven Press, 1990, vol 17, pp 227–287.
6 Hughes RAC: Protection of rats from experimental allergic encephalomyelitis with antiserum to guinea-pig spinal cord. Immunology 1974;26:703.

7 Galeone M, Moise G, Cacioli D, Megevand J, Bignamini AA: The use of antisera ('Serocytol') in the management of gastritis: a double-blind assessment versus placebo. Curr Med Res Opin 1985;9: 642–649.
8 Bahous I: Primäres Fibromyalgie-Syndrom. Bewertung der Wirksamkeit von Antigewebe-Immunoglobulinen (Ser 314) gegen Placebo in einer Doppelblindstudie. Therapiewoche Schweiz 1991;7: 710–720.
9 Kerkhofs M, Hoffman G, De Martelaere V, Linkowski P, Mendlewicz J: Sleep EEG recordings in depressive disorders. J Affect Disord 1985;9:47–53.
10 Carette S, Mc Cain GA, Bell DA, Fam AG: Evaluation of amitryptiline in primary fibrositis. Arthritis Rheum 1986;29:655–659.
11 Goldenberg DL, Felson DT, Dinerman H: A randomised controlled trial of amitriptyline and naproxen in the treatment of patients with fibromyalgia. Arthritis Rheum 1986;29:1371–1377.
12 Russel IJ, Fletcher EM, Michalek JE, McBroon PC, Hester GG: Treatment of primary fibrositis/fibromyalgia syndrome with ipubrofen and alprazolam. Arthritis Rheum 1991;34:552–560.
13 Kempenaers C, Simenon G, Vander Elst M, Fransolet L, Staner L, Appelboom T, Mendlewicz J: EEG sleep parameters in fibromyalgia. Sleep Res 1991;20A:91.
14 Reynolds WJ, Modolfsky H, Saskin P, Lue FA: The effects of cyclobenzaprine on sleep physiology and symptoms in patients with fibromyalgia. J Rheumatol 1991;18:452–454.
15 Adam K, Oswald I: Protein synthesis, bodily renewal and the sleep-wake cycle. Clin Sci 1983;65: 561–567.
16 Clot J, Andary M, Mingard P: In vitro and in vivo immunomodulating properties of horse antibodies to reticuloendothelial system (abstract). First World Conference on Inflammation, Antirheumatics, Analgesics and Immunomodulators, Venice, April 1984.

L. Staner, Sleep Laboratory, Department of Psychiatry, Erasme Hospital,
B–1070 Brussels (Belgium)

Basic and Clinical Aspects of Atopic Dermatitis

G. Rajka

Department of Dermatology, Rikshospitalet, Oslo, Norway

Atopic dermatitis (AD) is a common skin disease occurring in atopic persons, indicating a common heredity and a particular immune response, i.e. the production of IgE antibodies. Since varying skin lesions may be present in an atopic person it is essential to make a definition, which is described in table 1.

Concerning the prevalence, a trebling between 1945 and 1975 was observed [1], recent data from Japan [Uehara, pers. commun.] and Denmark [2] confirm this increase. In the global perspective, the disease is widespread, the prevalence is only low in India, Thailand and in some African states [3].

Already the definition reflects the primary role of itching which also can be proved experimentally. I showed that the duration of itch induced by the intracutaneous injection of trypsin, which mostly acts via histamine liberation taking an arbitrary limit of 2 min, is longer in AD than in controls. An argument of the role of itch in AD is given by a pathologist (table 2) [4]. The intense itching elicits scratching marks and after longer periods, at the same place, inflammatory changes follow such as prurigo papules and later on lichenification, i.e. inflammatory resonses. These are stictly correlated to the eczematous inflammation, a pillar of AD. The intense inflammation elicits itch and the ensuing scratching increases the eczematous inflammation, thus a vicious circle arises which is of great clinical significance. Whereas AD is basically a disease of infants and children, it may often occur in teenage and even in adult age.

The correlation between atopic skin, lung and nasal manifestations, i.e. between AD, asthma and hay fever is well documented [3].

Table 1. Definition of atopic dermatitis

Atopic dermatitis is a specific dermatitis in the abnormally reacting skin of the atopic person, resulting in itch with sequelae as well as in eczematous inflammation

Previously published in Rajka [3].

Table 2. On histopathology of AD

Atopic dermatitis is also largely caused by rubbing and scratching. If the fingers of a patient with atopic dermatitis could be restrained, the skin lesions would virtually disappear, although the itching would doubtless persist [4].

Previously published in Rajka [3].

Table 3. Allergens related to AD

Aeroallergens	mites molds animal hair pollen	Living agents	staphylococci (?) dermatophytes *Pityrosporon ovale*
Contact allergens	aeroallergens nickel, etc. (foods for children)	Food allergens	

Mechanism

The first major aspect of the mechanism in AD is related to the *atopic allergen*. By introducing different allergen extracts into the skin by intracutaneous or prick techniques, a large majority of AD patients with only skin affection ('pure' AD patients) have shown 80% positivity to a series of allergens [3]. It can also be shown that only a part of these antigens are allergens, i.e. capable of producing allergic reactions [6]. Technically, the prick test performed with a lancet seems to be appropriate, eliciting an immediate urticarial reaction and reading on average after 15 min.

The allergens playing major role in AD are shown in table 3. Of the aeroallergens, mites allergen can elicit in a larger proportion of AD patients, particularly in those over 21 years and having positive immediate reaction, a specific IgE antibody reaction and having most reaction in AD [7]. I have shown earlier that AD patients may have positive immediate reactions to several types of molds which also shows cross-reactivity to trichophytin, i.e. a dermatophyte extract [13]. The other aeroallergens are of minor significance in AD. On the other hand, mites can also act on the skin on the AD patient by contact. According to the concept of Bruijnzeel-Koomen [8] the mite allergen penetrates the epidermis, binds to the IgE of Langerhans cells which migrates to the dermis and activates T cells. The ensuing inflammatory delayed-type reaction makes the epidermis more permeable to further mite allergens which by binding to the IgE of mast cells initiates an immediate reaction.

Several allergens like chromium, rubber chemical, disinfectants, etc. may elicit allergic contact sensitivity in the AD patients, although to a lesser degree than in controls (see below) with the exception of nickel. Some proteins (such as food items or latex) elicit contact urticaria and contact with foods may lead to protein contact dermatitis (i.e. an eczematous reaction where also immediate response is found) [9]. Of living agents staphylococci play the leading clinical role. Their colonisation is under 10% on the normal skin, however 76% on symptom-free skin in AD and in 93% on lesional AD skin [10]. Dermatophytes and – in older teenagers and in adults – *Pityrosporon orbiculare* is a frequent colonisator (on the upper body half). In addition to colonisation, these microorganisms also elicit specific immune reactions. This is discussed concerning staphylococci whereas frequent immediate reactions to the dermatophyte extract trichophytin [3] occur and positive immediate reactions to *Pityrosporon orbiculare* in AD are also observed [11].

Food items are important allergens in AD, particularly at an early age, but it should be emphasized that food also can act by a nonallergenic (idiosyncrasic) mechanism [3].

The next major aspect is the *antibody production* in AD. Even if the IgE is not specific for AD and does not occur in every AD patient, this is the dominant antibody response in this disease. Elevated titers of IgE occur in several conditions and are quantified by a radioimmunosorbent technique (RIST). The specific antibody response to an allergen may be estimated by a radioimmunological in vitro assay [12]. In the field of AD there is a good correlation between RAST and skin test results with some minor exceptions.

Delayed Reactivity (Cell-Mediated Immunity)

I could already in 1963 demonstrate, by applying a series of microbial allergens (recall allergens), that the responses are greatly reduced in AD (table 4) [13]. This may be one of the major causes for the well-known higher susceptibility to microbial diseases (see below) and for the generally lower threshold to irritants as well as the generally reduced sensitization rate to contact allergens. The probable explanation is a reduction of T suppressor cells (OKT 8, now CD 8) versus normal number of T helper (OKT 4: CD4) cells thus changing their ratio, demonstrated in earlier experiments, at least in severe AD cases [14].

Table 4. Positive delayed reactions in atopic dermatitis and control groups

Allergens	I. Atopic dermatitis	II. Controls with skin diseases	III. Respiratory atopics
Streptococcal extract	3/50	27/50	15/25
Staphylococcal extract	5/50	29/50	14/25
Tuberculin 0.1 mg	0/50	13/50	5/25
PPD 0.002 mg/ml	11/16	16/16	
Mumps vaccine 1:10	8/16	15/16	
Schick test	6/16	10/16	
Trichophytin 1:50	1/30	7/20	

Previously published in Rajka [3].

Table 5. Immunological reactivity in AD

AD = immediate and delayed, possibly late reactions

Thus persons with AD easily/frequently get:

Immediate-type lesions	Contact urticaria Itch/edema in the mouth
Delayed-type lesions	Contact dermatitis Nummular eczema Seborrheic eczema
Immediate → delayed reactions	Inhalant contact dermatitis Protein contact dermatitis (Mucosal aphthae)
Immediate → late reactions, e.g. to food	

It is, however, obvious that the basis of the eczematous reaction in AD is a delayed-type reaction which is perhaps due to dyregulation of T helper cell subgroups. In addition, there are observations of IgE-bound late reactions in AD [15]. Thus, a spectrum of reactions is known in AD due to these immune mechanisms (table 5). Whereas the mediators of the immune reactions in AD are identical with other mast cell activating processes like urticaria, and the eczematous response does not differ (except for minor histological changes) from other ecze-

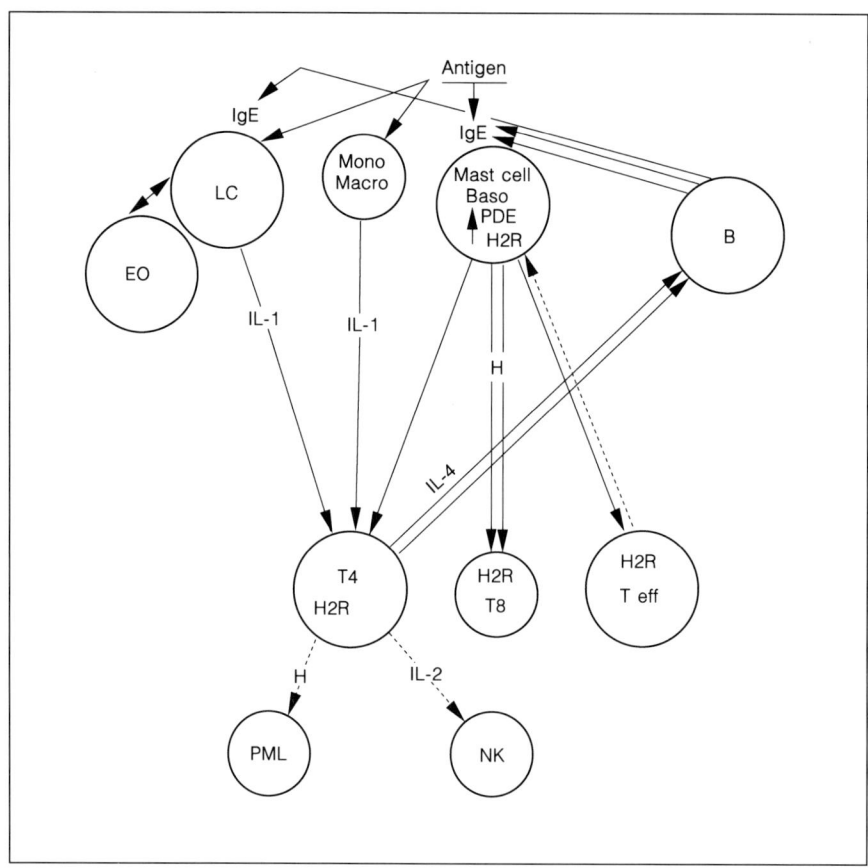

Fig. 1. Some schematically visualized data on relations between cells and mediators in AD. This figure has been published previously in Rajka [3].

matous diseases, there are particular biochemical alterations in AD concerning cyclic nucleotides [16].

I have depicted some major immunological events in a model. The consequence of the increased PDE/reduced cAMP activity is an increased mediator release affecting different T cells and indirectly polymorphonuclear and natural killer cells (even if by a feedback mechanism an opposite effect follows via H_2 receptors). Central to the mechanism is the production of IL-4 by T helper cells ultimately resulting in high IgE production which later on is deposited on the surface of monocytes, Langerhans and T cells (fig. 1). The recently proposed most exact mechanism is mediated by T helper cell subgroups, i.e. IL-4 produced by TH2 cells stimulating IgE production and inhibiting TH1 activity: delayed reac-

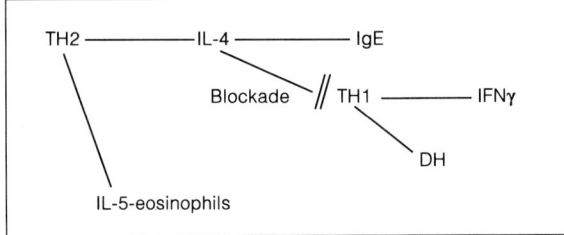

Fig. 2. Major cytokine profile for TH cells.

tivity and IFNγ production, which should usually inhibit the effect of IL-4 (fig. 2). In addition, IL-5 produced by TH2 cells activates eosinophils which seem to have major functions in AD: eosinophil major protein is elevated in AD skin, and eosinophil cationic protein serum level seems to occur parallel to the intensity of the disease [17].

Altered Skin

In my opinion, it is of basic significance that the immune processes in AD appear in a skin which is morphologically/functionally altered. Although the non-lesional skin of the AD patient macroscopically seems to be intact, it is altered in several aspects, therefore I speak of symptom-free skin. The list of alterations is shown in table 6, about which several remarks may be made. I have demonstrated a lower itch threshold in symptom-free AD skin, similarly to above-mentioned observations on lesional skin [3]. The consequences of the impaired barrier function are a lower threshold to irritants demonstrated by reactivity to cantharidin [3] and high staphylococcal colonisation. The characteristically dry skin may be explained by reduced sebaceous gland activity, particularly in women and by increased transepidermal water loss (TWL) [3]; in addition, a statistically proven combination with mild types of ichthyosis occurs. The paradoxical vasoconstrictive responses like white dermographism, altered nicotinate reaction and, following intracutaneous injection of acetylcholine, the delayed blanch phenomenon [18] are striking in the involved skin but may also occur in symptom-free skin.

In a further model I try to summarize the major aspects of the mechanism. The basic changes are the itch and correlated immune reactions (resulting in immediate and delayed, probably late responses) and the altered skin which also leads to itch, which thus can be considered as the central basic feature of the disease and I have tried to connect the different provoking factors in AD to these basic features (fig. 3).

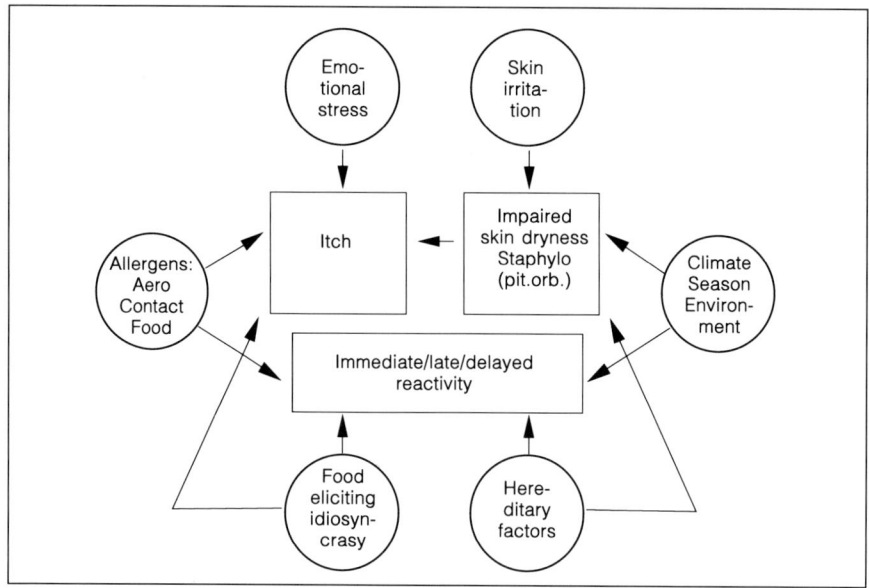

Fig. 3. The more important provocative factors in AD and their relation to the basic features.

Table 6. Alterations in symptom-free AD skin

1. Increased itchiness (lower itch threshold)
2. Increased epidermal thickness
3. Impaired barrier function
4. Lowered resistance to contact irritants
5. Increased staphylococcal colonization
6. Increased TWL
7. Coexistence with dry skin
8. Slight dermal lymphocyte infiltrate, alteration in venules
9. IgE on antigen-presenting cells[a]
10. White dermographism[b]
11. Nicotinate white reaction[b]
12. Delayed blanch[b]
13. Acral vasoconstriction
14. Absence of flare after histamine or allergen injection
15. Increased pilomotor reaction

[a] Found by one group.
[b] Not unanimously stated.
Previously published in Rajka [3].

Table 7. Guidelines for the diagnosis of AD

Must have 3 or more basic features:

Pruritus

Typical morphology and distribution
a Flexural lichenification or linearity in adults
b Facial and extensor involvement in infants and children

Chronic or chronically relapsing dermatitis
Personal or family history of atopy (asthma, allergic rhinitis, AD)

Minor features (3 or more)

General principle: Course influenced by environmental and emotional factors

I	Clinical signs	Early age at onset
		Xerosis/Ichthyosis/palmar hyperlinearity/keratosis pilaris
		Dennie-Morgan infraorbital fold
		Nonspecific hand and foot dermatitis
		Cheilitis
		Orbital darkening
		Anterior subcapsular cataract
		Pityriasis alba
		Nipple dermatitis
		Perifollicular accentuation
		Keratoconus
		Anterior neck folds
II	Immunological signs	Immediate skin test reactivity/elevated serum IgE
		Food intolerance
		Allergic conjunctivitis
		Impaired CMI/tendency toward skin infections
		(particularly *St. aureus,* herpes simplex)
III	Pathophysiological signs	Itch when sweating
		Intolerance to lipid solvents and wool
		Facial pallor/erythema
		White dermographism/delayed blanch

Although the diagnosis of AD appears to be quite straightforward, there are borderline cases and several other skin conditions in atopic individuals. Thus, it is of importance to establish diagnostic criteria. Our guidelines are as shown in table 7 [19, modified]. Estimating the severity of the disease, i.e. introducing a grading, we have proposed certain criteria [19].

Table 8. Generally recommended measures

1	Genetic counseling

2	Food avoidance
	A Breast feeding
	B Dietary measures in children/adults
	  i Strong food allergens (including contact with skin and mucosa) for clinically proven cases for a time period
	    Since it may be inconclusive in individual cases, it cannot be routinely indicated
	  ii Idiosyncrasy-eliciting items (in active phase of AD)

3	Avoidance of aeroallergens
	A Keeping pets/contact with horses, cows, pigs, etc.
	B Against mites/housedust, etc.
	  i Thorough cleaning
	  ii Removal of bedroom carpets, dust-attracting curtains, upholstered furniture
	  iii Use of polyester-filled pillows/covering of mattresses with impermeable fabric/frequent washing of bed clothes (with mild methods)
	  iv Acaricide chemicals
	C Against mold exposure (e.g. old houses, moist localities, storerooms, stables)
	D Against pollution (e.g. from local industries)
	  Smoking?
	E Against pollens (in some relevant cases); exclude cross-reacting food allergens

4	Against skin dryness (lubricants, bathing, bath oils, room humidification)
5	Occupational prophylaxis (including household work, hobbies)
6	Avoidance of overheating/sweating (work place, spiced food, exercise, etc.)
7	Avoidance of woollen/rough garments /'impermeable' textiles
8	Use of less-irritating diapers (e.g. hydrogel absorbent, paper) for infants/children with risk of or with manifest AD
9	Information for and registration of AD patients; parental education for infant/child AD patients

Table 9. Therapy

Present management

1	Systemic	Antihistaminics
		Steroids (short cures)
		Antibiotics
		(Acyclovir)
		(Ketokonazol)
2	Topical	Antipruritics
		Steroids
		Tar
		Antibacterial agents
		Lubricants

Newer management

1	Systemic	Cyclosporin-A
		Loratadin/Cetirizin
		UV AB/UV A1/narrow band
2	Topical	Mometason furoat
		Duoderm + clobetasol prop.
3	Experimental	Essential fatty acids: n–3 (n–6) series
		Thymopentin
		Interferon-α
		Hyposensitivity with
		D. pteronyssimus complexes
		IL-1?

From the point of view of prophylaxis the major events are collected in table 8. Finally, I summarize the most used present therapeutic principles and the newer aspects of management of AD (table 9).

References

1   Taylor B, Wadworth M, Wadworth J, Peckham C: Changes in the reported prevalence in childhood eczema since the 1939–1945 war. Lancet 1984;i:1255.
2   Schultz Larsen F, Hanifin JM: Secular change in the occurence of atopic dermatitis. Acta Derm Venereol (Stockh) 1992;(suppl 176):7.
3   Rajka G: The Essential Aspects of Atopic Dermatitis. Berlin, Springer, 1989.
4   Ackerman B: Histologic Diagnosis of Inflammatory Diseases. Philadelphia, Lea & Febiger, 1978, p 259.

5 Bono J, Levitt P: Relationship of infantile atopic dermatitis to asthma and other respiratory allergies. Ann Allergy 1964;22:72.
6 Løwenstein H, Lind P, Wecke B: Identification and clinical significance of allergenic molecules of cat origin. Part of the DAS/76 study. Allergy 1985;40:430.
7 Chapman MD, Rowntree S, Mitchell EB, Di Prisco Fuenmayor MC, Platts-Mills TAE: Quantitative assessment of IgG and IgE antibodies to inhalant allergens in patients with atopic dermatitis. J Allergy Clin Immunol 1983;72:27.
8 Bruijnzeel-Koomen C: IgE on Langerhans' cells: New insight into the pathogenesis of atopic dermatitis. Dermatologica 1986;172:181.
9 Hjorth N, Roed-Petersen J: Occupational protein dermatitis in food handlers. Contact Derm 1976; 2:28.
10 Aly R: Bacteriology of atopic dermatitis. Acta Derm Venereol (Stockh) 1980;(suppl 92):16.
11 Waersted A, Hjorth N: *Pityrosporon orbiculare.* A pathogenetic factor in atopic dermatitis of the face, scalp and neck? Acta Derm Venereol (Stockh) 1985;(suppl 114):146.
12 Wide L, Bennich H, Johansson SGO: Diagnosis of allergy by an in vitro test for allergen antibodies. Lancet 1967;ii:1105.
13 Rajka G: Studies in hypersensitivity to molds and staphylococci in prurigo Besnier (atopic dermatitis). Acta Derm Venereol (Stockh) 1963:(suppl 43).
14 Braathen LR: T cell subsets in patients with mild and severe atopic dermatitis. Acta Derm Venereol (Stockh) 1985;(suppl 114):133.
15 Sampson HA: The role of food allergy and mediator release in atopic dermatitis. J Allergy Clin Immunol 1988;81:635.
16 Hanifin JM: Pharmacophysiology of atopic dermatitis. Clin Rev Allergy 1986;4:43.
17 Czech W, Krutmann J, Schöpf E, Kapp A: Serum eosinophil cationic protein (ECP) is a sensitive measure for disease activity in atopic dermatitis. Br J Dermatol 1992;125:351.
18 Lobitz WC, Campbell: Physiologic studies in atopic dermatitis (disseminated neurodermatitis). I. The local cutaneous response to intradermally injected acetylcholine and epinephrine. Arch Dermatol 1953;67:575.
19 Hanifin JM; Rajka G: Diagnostic features of atopic dermatitis. Acta Derm Venereol (Stockh) 1980; (suppl 92):44.

G. Rajka, Department of Dermatology, Rikshospitalet, N–0027 Oslo 1 (Norway)

The Actions of Antihistamines in Allergic Disease

Martin K. Church

Clinical Pharmacology Group, Southampton General Hospital, Southampton, UK

Allergic diseases comprise a spectrum of phathological conditions which share a common initiation and have a remarkably similar cellular pathology and yet whose symptoms vary considerably with the nature of the target organ. In a sensitized individual, the primary event following allergen challenge of a mucosal membrane is activation of tissue mast cells by cross-linkage of their membrane-bound IgE. This results in the generation into the extracellular environment of both preformed and newly generated mediators. Of the preformed mediators released by exocytosis, the vasoactive amine, histamine, is the best characterized, its actions on the local environment to causing vasodilation, plasma extravasation and smooth muscle contraction, particularly in the airways and intestine. Stimulation of phospholipid metabolism in the mast cell membrane generates significant quantities of prostaglandin D_2 (PGD_2), leukotriene C_4 (LTC_4) and platelet-activating factor (PAF). The former two products are potent contractors of bronchial smooth muscle while PAF has been suggested to be an initiator of inflammation and to induce bronchial hyperreactivity. The recent identification of cytokine interleukins, particularly IL-4, IL-5 and IL-6 and TNF-α as mast cell products [1, 2] suggests that the mast cell also has a role in the initiation and maintenance of allergic inflammation.

The local up-regulation of selectins, adhesion proteins and chemotactic factors initiated by the early phase of the allergic response leads to an initial influx of neutrophils followed by a slower and more prolonged accumulation of eosinophils, the latter being characteristic of allergic inflammation. The extracellular actions of the eosinophil-derived proteins, major basic protein (MBP), eosinophil cationic protein (ECP) and eosinophil peroxidase (EPO), and of the newly generated mediators PAF, LTC$_4$ and oxygen free radicals have led to the association of the eosinophil with the late-phase allergic response and chronic allergic inflammation.

Antihistamines as Inhibitors of Early Phase Allergic Responses

At least two factors influence the ability of a histamine H$_1$ receptor antagonist to ameliorate a disease process: the relative contribution of histamine to the aspect of the disease under consideration and the pharmacological effectiveness of the drug in antagonizing the receptor-mediated effects of histamine in the clinical environment.

The relative contribution of histamine to the symptoms or progression of any particular allergic disease will itself depend on two factors: the relative concentrations of mediators derived from the mast cell and the relative sensitivity of the target cell for these mediators. In the early phase of the allergic response, mast cell activation generates into the extracellular environment histamine, PGD$_2$ and LTC$_4$ in approximate ratios of 1,000:25:2 [3]. Target tissues vary considerably in their sensitivity to these biologically active substances. For example, histamine is a more effective vasodilator and inducer of vascular leakage than is PGD$_2$ or LTC$_4$. As a consequence, allergic conditions where vascular responses are major contributors to the symptoms such as rhinorrhea or urticaria, histamine H$_1$ antagonists have the potential for being very effective drugs. In contrast, the effectiveness of histamine, PDG$_2$ and LTC$_4$ in causing bronchoconstriction is approximately the inverse of the amounts released from the mast cell. As a consequence, all mediators contribute approximately equally to the early-phase allergic bronchoconstriction and histamine H$_1$ antagonists will be, at best, only partially inhibitory.

The second criterion for drug activity is its pharmacological effectiveness in the clinical situation. Histamine H$_1$ antagonists, unlike H$_2$ antagonists, show little structural similarity to histamine but resemble more closely antimuscarinic and antidepressant drugs. This means that the chemical molecule contains the potential for diverse pharmacological activities, many of which may be unwanted [4]. Indeed, many of the older antihistamines, including mepyramine, diphenhydramine, tripelennamine, chlorpheniramine and promethazine, have antagonistic effects at muscarinic, cholinergic, α-adrenergic and serotoninergic receptors, act as local anesthetics, and, perhaps most importantly, readily cross the blood-brain barrier to cause central nervous system (CNS) sedation. Direct impairment of psychomotor performance and potentiation of the sedative effects of alcohol and CNS depressants, such as minor tranquillizers, have severely limited the use of these drugs. Over the last decade, the problem of CNS sedation with histamine

H_1 receptor antagonists has been significantly reduced by the introduction of a new generation of compounds, including terfenadine, astemizole, cetirizine, loratadine, azelastine, picumast, acrivastine and mequitazine. We must not be complacent, however, for newer side effects are emerging with high doses of some antihistamines. For example, retardation of the metabolism of terfenadine by coadministration of erythromycin or ketoconazole results in prolongation of the cardiac Q-T interval leading to torsades des pointes in susceptible individuals [5].

With the older antihistamines, only doses capable of affording up to 3 to 5-fold protection against histamine-induced bronchoconstriction may be given before CNS sedation becomes a problem. In contrast, the newer generation of antihistamines may be given at doses which afford 30- to 40-fold protection against histamine under the same conditions before unwanted effects become apparent. Thus, although both generations of drugs may be used successfully to treat urticaria and rhinitis, some sedation will be obvious with the older antihistamines even though a relatively low degree of H_1 receptor blockade is necessary to improve symptoms.

The usefulness of antihistamines in the treatment of bronchial asthma is, however, far less clear cut because of the wide diversity of pharmacological mediators released and the multiplicity of end organ effects which go to make up this clincal syndrome. The potential involvement of histamine is evidenced from two sets of data. First is the observation of elevated plasma histamine levels in stable asthma [6] consistent with which is the long recognized weak bronchodilator actions of both the older and newer antihistamines [7, 8]. Second are the observations under experimental conditions that elevated plasma histamine levels accompany bronchoconstriction following provocation with allergen [9]. However, as eicosanoids also contribute significantly to allergen-induced bronchoconstriction, this condition is only partially abrogated by antihistamines [10]. With the availability of β-adrenoceptor stimulants and theophylline-like drugs, antihistamines are rarely used for symptomatic relief in acute bronchoconstriction.

Nonhistamine H_1 Receptor-Dependent Effects of Antihistamines

It is becoming clear that while the development of highly specific histamine H_1 receptor antagonists for the treatment of some allergic conditions is becoming a reality, when it comes to the treatment of bronchial asthma then a drug which possesses additional pharmacological properites may well be more effective. Such additional properties would include mast cell stabilization, inhibition of eosinophil accumulation and activation, inhibition of the release and actions of preformed and newly generated mediators from inlammatory cells and up regulation

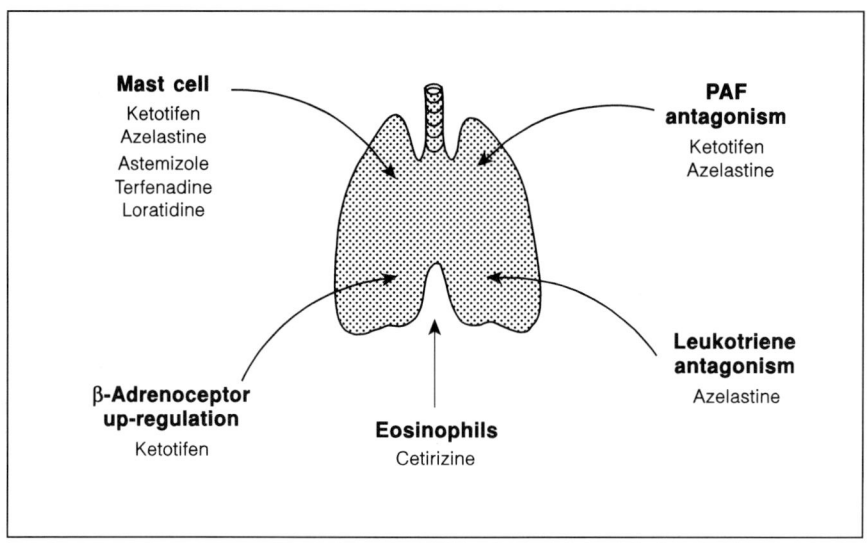

Fig. 1. Purported anti-inflammatory effects of antihistamines.

of β-adrenoceptor coupling (fig. 1). Furthermore, not only do these actions have to be clearly demonstrable in an experimental situation, but they must occur at concentrations likely to be achieved during treatment in the clinical setting. The purported abilities of antihistamines to prevent mast cell mediator release and to reduce eosinophil accumulation and activation are reviewed.

Antihistamines and Mast Cell Mediator Release

The ability of antihistamines to influence mast cell mediator release has been recognized for many years [11, 12]. At lower concentrations, antihistamines inhibit histamine release whereas at higher concentrations they induce release, neither effect appearing to be related to the ability of the drugs to interact with histamine H_1-receptors [13]. Structure-activity studies [13] have suggested that two physical characteristics, lipophilicity and net positive charge are the major determinants of these effects. The high dose release of histamine has been suggested to result from disruption of the cell membrane following physical incorporation of the drug in the mast cell membrane [14]. The purported mechanisms by which antihistamines prevent mast cell mediator release are less well understood. Mechanisms proposed include a nonspecific membrane stabilizing effect [14] and an interference with calcium mobilization either within the membrane [15] or at the level of calmodulin [16]. In the in vitro situation at least, the concentrations of drugs required to inhibit histamine release from mast cells and basophils are in the micromolar range [16–18].

Evidence for whether or not mast cell stabilization is a clinically relevant phenomenon is controversial. The observations that the IC_{50} values for the inhibition of histamine release from lung mast cells by azelastine is 3 µM [19] and from skin mast cells by terfenadine is 2–10 µM [Okayama and Church, unpubl. observations] whereas the peak blood levels of these drugs achieved during therapy are only 3–30 nM would suggest little activity in man. However, evidence for clinical activity has been reported, often using doses of drugs 2–5 times higher than those normally used in clinical practice. In the skin chamber model, hydroxyzine significantly reduces allergen-induced mast cell degranulation [20] while cetirizine variably inhibits PGD_2 release but not that of histamine [21]. Using a nasal challenge model, terfenadine [22] and loratadine [22], but not cetirizine [23], diphenhydramine [24] or ketotifen [25], have been reported to inhibit mast cell histamine release following allergen provocation. In the lower airways, studies designed to explore reduction of histamine release have, to date, proved negative, neither astemizole [26] nor azelastine [27] inhibiting the rise in plasma histamine levels while ketotifen also failed to prevent the rise in serum neutrophil chemotactic activity following allergen provocation [28]. Thus, it may be concluded that inhibition of mast cell mediator release by antihistamines in the clinical situation may not be a ubiquitous phenomenon but could contribute to the beneficial effects of some drugs under selected conditions.

Antihistamines and the Accumulation and Activation of Inflammatory Cells

The accumulation and activation of inflammatory cells underlies the whole process of allergic inflammation and the ability of a drug to prevent this would be a great attribute for the treatment of the chronic aspects of allergic disease.

Using the Boyden chamber to assess in vitro activity, cetirizine reduces eosinophil chemotaxis at concentrations well within the therapeutic range [29, 30]. In similar studies, dexchlorpheniramine, loratadine and ketotifen were either weakly active or required concentrations in the micromolar range to be effective [29, 31]. In vivo, local allergen challenge of the skin, nose and lungs has been used to assess the effects of drugs on eosinophil accumulation. Using the skin window technique the older antihistamines, chlorpheniramine and promethazine do not prevent eosinophil accumulation whereas corticosteroids are potently active [32]. Of the newer antihistamines, cetirizine has been shown repeatedly to reduce eosinophil accumulation in the skin following allergen challenge [21]. Using skin biopsies rather than a skin window, cetirizine was shown to have a weak effect which just failed to reach statistical significance [32] while astemizole was without effect [33]. Local allergen challenge of the nose has provided variable results, one study [34] showing both terfenadine and cetirizine to be inactive whereas another

[35] showed complete inhibition of eosinophilia. Similarly, in seasonal rhinitis in season, no reduction in eosinophilia has been reported by one group [36] whereas another reported a preferential inhibition of eosinophilia by cetirizine when compared with terfenadine or astemizole [37]. In asthma, only one study [38] using rather a high dose of cetirizine (15 mg twice daily) has appeared in the literature thus far. This study showed a significant reduction in pulmonary eosinophilia but, interestingly, no reduction in the symptoms of the late phase response as measured by airways function testing. Thus, as with mast cell stabilization, the ability of antihistamines to prevent eosinophil accumulation and activation is not uncontroversial and the activity of cetirizine requires optimisation.

Conclusions

With the newer generation of antihistamines, adequate antagonism of the end-organ effects of histamine with little or no CNS depression has now been attained. However, the quest to reveal additional beneficial actions, such as mast cell stabilization or inhibition of eosinophil accumulation, has meant that doses are being pushed higher and higher thus increasing the possibility of unwanted toxicity. Hopefully, molecular manipulation and targeted research will, in the near future, lead to the development of antihistaminic drugs with even greater anti-allergic effects and a high therapeutic index.

References

1 Bradding P, Feather IH, Howarth PH, Mueller R, Roberts JA, Britten K, Bews JPA, Hunt TC, Okayama Y, Heusser CH, Bullock GR, Church MK, Holgate ST: Interleukin 4 is localized to and released by human mast cells. J Exp Med 1992;176:1381–1386.
2 Gordon JR, Burd PR, Galli SJ: Mast cells as a source of multifunctional cytokines. Immunol Today 1990;11:458–464.
3 Robinson C, Benyon RC, Holgate ST, Church MK: The IgE- and calcium-dependent release of eicosanoids and histamine from human cutaneous mast cells. J Invest Dermatol 1989;93:397–404.
4 Simons FER, Simons KJ: Antihistamines; in Middleton E, Reed CE, Ellis EF, Adkinson NF, Yunginger JW, Busse WW (eds): Allergy: Principles and Practice. St. Louis, Mosby, 1993, pp 856–892.
5 Woosley RL, Chen Y, Freiman JP, Gillis RA: Mechanism of the cardiotoxic actions of terfenadine. JAMA 1993;269:1532–1536.
6 Barnes PJ, Ind PW, Brown MJ: Plasma histamine and catecholamines in stable asthmatic subjects. Clin Sci 1982;62:661–665.
7 Popa VT: Bronchodilating activity of an H1-blocker, chlorpheniramine. J Allergy Clin Immunol 1977;59:54–63.
8 Nogrady SG, Hartley JPR, Handslip PDJ, Hurst NP: Bronchodilation after inhalation of the antihistamine clemastine. Thorax 1978;33:479–482.

9   Howarth PH, Durham SR, Lee TH, Kay AB, Church MK, Holgate ST: Influence of albuterol, cromolyn sodium and ipratropium bromide on the airway and circulating mediator responses to antigen bronchial provocation in asthma. Am Rev Respir Dis 1985;132:986–992.
10 Rafferty P, Beasley CRW, Southgate P, Holgate ST: The role of histamine in allergen and adenosine-induced bronchoconstriction. Int Arch Allergy Appl Immunol 1987;82:292–294.
11 Arunlakshana O, Schild HO: Histamine release by antihistamines. J Physiol (Lond) 1953;119:47P–48P.
12 Mota I, Dias da Silva W: The antianaphylactic and histamine releasing properties of the antihistamines: Their effect on mast cells. Br J Pharmacol 1960;15:396–404.
13 Church MK, Gradidge CF: Inhibition of histamine release from human lung in vitro by antihistamines and related drugs. Br J Pharmacol 1980;69:663–667.
14 Seeman P: The membrane actions of anaesthetics and tranquillizers. Pharmacol Rev 1972;24:583–655.
15 Tasaka K, Mio M, Okamoto M: Intracellular calcium release induced by histamine releasers and its inhibition by antiallergic drugs. Ann Allergy 1986;56:464–469.
16 Peachell PT, Pearce FL: Divalent cation dependence of the inhibition by phenothiazines of mediator release from mast cells. Br J Pharmacol 1989;97:547–555.
17 Okayama Y, Church MK: Comparison of the modulatory effect of ketotifen, sodium cromoglycate, procaterol and salbutamol in human skin, lung and tonsil mast cells. Int Arch Allergy Appl Immunol 1992;97:216–225.
18 Rimmer SJ, Church MK: The pharmacology and mechanism of action of histamine H1-antagonists. Clin Allergy 1990;20 (suppl 2):3–17.
19 Little MM, Wood D, Casale TB: Azelastine inhibits stimulated histamine release from human lung tissue but does not alter cyclic nucleotide content. Am Rev Respir Dis 1987;135:317.
20 Ting S, Rauls DO, Reiman BEF: Inhibitory effect of hydroxyzine on antigen-induced histamine release in vivo. J Allergy Clin Immunol 1985;75:63–66.
21 Charlesworth EN, Kagey-Sobotka A, Norman PS, Lichtenstein LM: Effects of cetirizine on mast cell mediator release and cellular traffic during the cutaneous late phase response. J Allergy Clin Immunol 1989;83:905–912.
22 Bousquet J, Lebel B, Chanal I, Morel A, Michel FB: Antiallergic activity of loratidine and terfenadine assessed by nasal challenge. J Allergy Clin Immunol 1988;81:228.
23 Naclerio RM, Proud D, Kagey-Sobotka A, Friedhoff L, Norman PS, Lichtenstein LM: The effect of cetirizine on the early allergic response. Laryngoscope 1989;99:596–599.
24 Majchel AM, Proud D, Kagey-Sobotka A, Lichtenstein LM, Witek TJ, Naclerio RM: Persistent efficacy of a combination antihistamine/analgesic/decongestant product the morning after a bedtime dose. J Allergy Clin Immunol 1991;87:151.
25 Majchel AM, Proud D, Kagey-Sobotka A, Lichtenstein LM, Naclerio RM: Ketotifen reduces sneezing but not histamine release following nasal challenge with antigen. Clin Exp Allergy 1990;20:701–705.
26 Holgate ST, Emanuel MB, Howarth PH: Astemizole and other H1-antihistamine drug treatment of asthma. J Allergy Clin Immunol 1985;76:375–382.
27 Rafferty P, Ng WH, Phillips GD, Clough J, Church MK, Aurich R, Ollier S, Holgate ST: The inhibitory actions of azelastine hydrochloride on the early and late bronchoconstrictor responses to inhaled allergen in atopic asthma. J Allergy Clin Immunol 1989;84:649–657.
28 Morgan DJR, Moodley I, Cundell DR, Sheinman BD, Smart W, Davies RJ: Circulating histamine and neutrophil chemotactic activity during allergen induced asthma: The effect of inhaled antihistamines and anti allergic compounds. Clin Sci 1985;69:63–69.
29 De Vos C: H1-antagonists and inhibitors of eosinophil accumulation. Clin Exp Allergy 1991;21:277–281.
30 Townley RG, Okada C: Use of cetirizine to investigate non-H_1 effects of second-generation antihistamines. Ann Allergy 1992;68:190–196.
31 Sugiyama H, Nabe M, Miyagawa H, Agrawal DK, Townley RG: Effect of ketotifen on calcium ionophore induced LTC4 and platelet activating factor induced eosinophil chemotaxis. Am Rev Respir Dis 1990;141:A874.

32 Varney VA, Gaga M, Frew AJ, De Vos C, Kay AB: The effect of a single oral dose of prednisolone or cetirizine on inflammatory cells infiltrating allergen-induced cutaneous late-phase reactions in atopic subjects. Clin Exp Allergy 1992;22:43–49.
33 Bierman CW, Maxwell D, Rytina E, Emanuel MB, Lee TH: Effect of H1-receptor blockade on the late cutaneous reactions to antigen: A double-blind controlled study. J Allergy Clin Immunol 1991; 87:1013–1019.
34 Klementsson H, Andersson M, Pipkorn U: Allergen-induced increase in nonspecific nasal reactivity is blocked by antihistamines without a clear cut relationship to eosinophil influx. J Allergy Clin Immunol 1990;86:466–472.
35 Leonhardt L, Molitor SJ, Richter K: Delayed reactions and eosinophilia after nasal challenge test with allergens. Schweiz Med Wochenschr 1991;121:15.
36 Howarth PH, Wilson SJ, Brewster H: The influence of cetirizine on symptom generation and nasal eosinophilia in seasonal allergic rhinitis. J Allergy Clin Immunol 1991;87:151.
37 Spyropoulou M, Emmanuel B, Tsangarakis E, Stauropoulos-Giokas A: Comparative efficacy of non-sedating anti-histamines in seasonal allergic rhinitis. Clin Exp Allergy 1990;20 (suppl 1):56.
38 Rédier H, Chanez P, De Vos C, Rifaï N, Clauzel A-M, Michel F-B, Godard P: Inhibitory effect of cetirizine on the bronchial eosinophil recruitment induced by allergen inhalation challenge in allergic patients with asthma. J Allergy Clin Immunol 1992;90:215–224.

Prof. M.K. Church, Clinical Pharmacology Group, Centre Block, Southampton General Hospital, Tremona Road, Southampton SO9 4XY (UK)

New Targets for Antiallergic Agents

G. Ciprandi, S. Buscaglia, C. Pronzato, V. Ricca, G.P. Pesce, B. Villaggio, N. Fiorino, M. Albano, A. Scordamaglia, M. Bagnasco, G.W. Canonica

Allergy and Clinical Immunology Service, Department of Internal Medicine, DIMI, University of Genoa, Italy

Introduction

Inflammation is the response of vascularised tissue to injury and is usually beneficial since it serves to resolve and to repair the effect of damage. Several factors can initiate inflammation including infectious agents (bacteria, viruses and parasites), physical agents (burns, radiation and trauma), chemical agents (drugs, toxins and industrial agents), ischemic tissue injury and immunological disorders, such as, for example, allergy and autoimmunity [1, 2].

Particularly, the allergic reaction is characterized by a cascade of events beginning with the release of vasoactive, chemotactic and spasmogenic mediators and followed by an influx of inflammatory cells, i.e. neutrophils, eosinophils, mastcells, basophils, mononuclear phagocytes, platelets and lymphocytes into the site of tissue damage [2–4]. Activated inflammatory cells release a wide range of inflammatory mediators that include chemotactic factors, cytokines, preformed and newly formed mediators. Of particular interest is the induction/shedding and enhanced expression of cell-surface molecules on vascular endothelium and circulating leukocytes that leads to leukocyte margination, activation and subsequent migration out of the vascular space and into the inflamed tissue [5–8]. Cytokines, such as IL-1, TNFα or IFNγ can induce synthesis and expression of specific adhesion molecules on immunocompetent cells and target cells [9–12].

In addition, the epithelium is a dynamic tissue that may also play a role during allergic inflammation in the defence as well as in the damage of the target organs [13, 14]. For instance, in asthma and in other inflammatory diseases of the airways there are conspicuous alterations in the epithelium ranging from cell desquamation to fibrogenesis, many of which may result from the inflammatory process itself [2, 15].

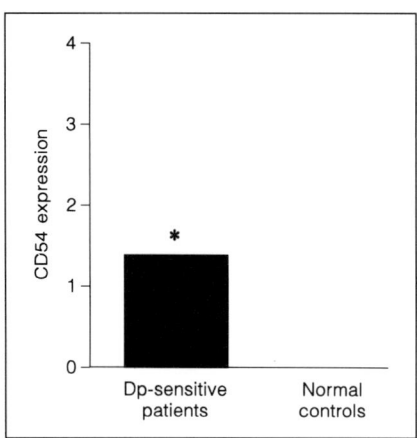

Fig. 1. Evidence of CD54 expression on conjunctival epithelium in basal condition in patients with allergy to house dust mites. * Significant vs. controls.

Recent studies investigated extensively the role of adhesion molecules in the allergic inflammation and one of the favourite topics has been the CD54 (ICAM-1)/LFA1 system [16–19].

In humans, first evidence of CD54 expression on epithelial cells of asthmatic subjects has been recently provided by Campbell at al. [15] who also demonstrated an enhanced expression when histamine is added to the cultured biopsies.

A chronic airway inflammation can be detected in all asthmatic patients, ranging from a mild degree in symptom-free individuals (so-called minimal persistent inflammation') to severe damage and bronchial epithelium shedding [2]. The presence of a mild inflammatory infiltrate of activated cells, such as neutrophils, eosinophils and lymphocytes as a constant feature in completely symptom-free subjects is worthy of note. Convincing evidence suggests a strong correlation between the degree of local inflammation and the level of bronchial hyperactivity, in other words, the minimal persistent inflammation renders the asthmatic individual more susceptible to exacerbation of the disease, although symptom-free [2, 15].

We have investigated the possible existence and features of a minimal persistent inflammation in other target organs of allergic reactions, such as the conjunctiva [in preparation]. The results obtained are consistent with those found in asthma: conjunctival scrapings have been obtained from patients suffering from asthma due to house dust mite, their relatives and healthy volunteers [20]; it is of

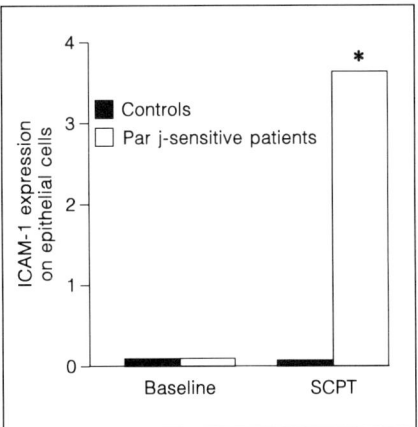

Fig. 2. Evidence of CD54 expression on conjunctival epithelium after CPT in allergic individuals and healthy volunteers. * Significant vs. controls. Modified from Ciprandi et al. [24].

note that allergic patients did not present with any conjunctival signs or symptoms on admission (fig. 1). Interestingly, a mild inflammatory cell infiltrate has been detected in mite-sensitive patients in basal conditions. Together with the cell infiltrate, mainly constituted by neutrophils, eosinophils and lymphocytes, CD54 expression on conjunctival epithelium was seen. No sign of inflammation appeared either in the patients' relatives (living in the same environment) or in the control group [20]. Moreover, the dust mite evaluation revealed that all houses of allergic patients contained a mite level sufficient to determine sensitization and inflammation, according to the literature [21]. This experimental evidence points out that also in situations of persistent exposure to allergens, other than mites, a minimal persistent inflammation can be detected even in clinical latency. This finding outlines the importance of the assumption that allergic asthma is a chronic inflammatory disease, which may be complicated by the development and maintenance of the inflammatory process by irreversible phenomena, i.e. subepithelial fibrosis and remodelling of the original structure of the target organ [2, 15]. Therefore, from a therapeutical viewpoint, the management of mite-sensitive patients should be based on preventive environmental measures to reduce allergen exposure and on precocious and possibly constant anti-inflammatory drug treatment such as cromones, corticosteroids and perhaps antiallergic drugs as well at low and effective dosages [22, 23].

We performed a conjunctival challenge in symptom-free patients with allergy due to pollen *(Parietaria judaica)* and healthy volunteers to evaluate the

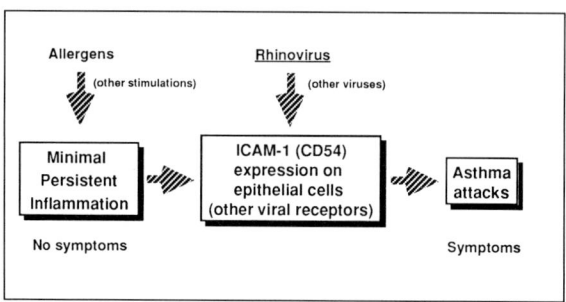

Fig. 3. Possible pathway for a viral induction of asthma.

clinical response and cellular reaction. The test was performed out of the pollen season [24]. Our data demonstrated CD54 expression on conjunctival epithelium after conjunctival challenge in allergic individuals (fig. 2). CD54 expression occurred as an early event persisting during the whole development of the allergic reaction, i.e. during late-phase reaction (LPR), vanishing 24 h after the challenge. The occurrence of this phenomenon following allergen exposure was specifically restricted to allergic subjects [24]. It is of note that such kinetic of appearance is very rapid compared to the kinetic of appearance of CD54, for instance, on lymphocytes following mitogen stimulation. In addition, we obtained experimental evidence that an identical phenomenon occurs on nasal epithelial cells following allergen-specific stimulation with the same kinetics [in preparation]. Moreover, our preliminary data have shown the expression of LFA-1 on lymphocytes simultaneously with CD54 expression on conjunctival epithelial cell after challenge [in preparation].

The changes in surface adhesion molecule pattern expressed by inflamed epithelium of bronchial mucosa in asthma, or other target organs of allergic reactions, specifically the expression of CD54, is also relevant as far as the susceptibility to viral infections is concerned. In fact, CD54 has been shown to bind rhinoviruses: about 90% of rhinoviruses use CD54 as a receptor, their binding site being distinct from, although overlapping LFA-1 [25–27]. Therefore, a possible causal link can be hypothesized between exacerbation of asthma and viruses, as recently appreciated in several epidemiological studies (fig. 3) [28]. At this stage it might be hypothesized that the demonstrated minimal persistent inflammation is capable of inducing a discrete CD54 expression on respiratory epithelium which may become susceptible to viruses binding.

Finally, the central role of endothelium in the migration of inflammatory cells and localization into the target organs should be underlined: the induction of endothelial leukocyte adhesion molecules represents a fundamental and crucial event in the accumulation of inflammatory cells at the site of antigen-stimulated tissue. ELAM-1 antigen occurs concurrently with the development of inflammatory cell infiltrates during late-phase allergic skin reactions [13, 14]. Furthermore, the local secretion of IL-1 and TNF are required for the induction of ELAM-1 expression [12].

A next series of studies was designed to examine the effects of drugs employed commonly in the treatment of allergic diseases (i.e. antiallergic drugs) on clinical and cytological events. We first evaluated terfenadine and loratadine, both new third-generation H1 antagonists. Terfenadine showed a significant protective effect on the early-phase cellular and clinical events of conjunctival reaction induced by allergen challenge in atopic patients. This effect did not appear dose-dependent [29]. Loratadine protected against the early clinical and cellular changes following the conjunctival provocation test (CPT) [30]. More recently, we proved the protective effects of loratadine also on the late-phase events consequent to CPT [31]. With regard to all these data, we studied Deflazacort, a new corticosteroid compound, and cetirizine, new third-generation antihistamine compound. The studies were designed to investigate the possible effects of both drugs on conjunctival clinical and cytological changes (including CD54 expression on conjunctival epithelium) occurring after CPT in sensitized patients.

Materials and Methods

Drugs

Deflazacort, an oxazoline derivative of prednisolone, is a new synthetic corticosteroid agent with calcium and glucose-sparing and bone-saving activities [32]. Its anti-inflammatory activity has been shown both in in vitro [32–34] and in in vivo models [32, 33].

Cetirizine, carboxylated metabolite of hydroxizine, is a potent histamine H_1 receptor antagonist, lacking the central nervous system depressant effects of standard analogous antihistamines, and shows a clinical effectiveness when employed in treating rhinoconjunctivitis and allergic urticaria [35]. Cetirizine exerts antiallergic activity preventing clinical changes consequent to allergen-specific challenge perfomed at the conjunctival, nasal and cutaneous levels. Moreover, cetirizine reduces in vivo eosinophil and neutrophil infiltration to the site of allergic reaction and platelet-activating factor (PAF) release and in vitro eosinophil chemotaxis to PAF [35]. In contrast, cetirizine does not inhibit in vivo histamine release by mast cells and in vitro lymphocyte proliferation and surface molecule expression [34, 35].

Model

We employed the CPT with allergens, since it is a fruitful in vivo model to assess pharmacological modulation of allergic response [29–31, 36]. CPT evaluation may be performed by clinical examination, cytological evaluation and/or analysis of mediator release in the lachrymal fluid [4, 37]. Moreover, we have recently used a sensitive immunocytochemical APAAP technique to detect the possible expression of the CD54 molecule on conjunctival epithelium in allergic subjects as an additional parameter to study the allergic reaction [24, 38].

The exposure of the conjunctival mucosa to specific allergen results in an immediate hypersensitivity reaction occurring 20–30 min after challenge and quickly subsiding. This early-phase reaction (EPR), similar to allergen provocation in the skin [39], lung [40] and nose [41], is characterised by clinical symptomatology and local infiltration of inflammatory cells, mainly neutrophils [4]. Bonini and co-workers and our group described a second clinical reaction, termed the late-phase reaction (LPR), appearing 1 h later, peaking at 6–8 h after exposure, and slowly (24 h more) subsiding. LPR shows a predominance of eosinophils and presence of lymphocytes [4, 24, 31, 36, 42]. No quiescent (symptomless) pause exists at the conjunctival level; instead, a 'continuous' reaction is observed. This evidence is likely to be due to the high allergen dose employed in CPT, compared to other organs, such as the nose, bronchus and skin [39–41]. Both clinical evidence and cellular infiltration may vary according to allergen dose employed. At high allergen doses a constant cellular late phase and a frequent clinical late phase dominate [4, 31, 42], whereas only a cellular infiltrate is observed at low allergen doses [4, 24, 29, 30].

Design

Both studies were placebo-controlled, double-blind and randomized. A preliminary CPT ('screening test') was performed weekly, for 3 weeks, to screen patients positive for LPR with *Parietaria judaica* extract at different dilutions. After screening, selected patients (those who showed LPR) were included in the trials and randomized to double-blind treatment [31, 42]. The trials were performed in the off-pollen season (December–February), during remission of symptoms.

Deflazacort trial included 24 of 29 patients (those who showed LPR during a screening test). They were randomly assigned to four parallel groups of treatment, receiving an oral daily dose of Deflazacort at 6, 30 and 60 mg/die or placebo, in the morning for 3 days [42]. Cetirizine trial included 12 of 15 patients (those who showed LPR during a screening test). They were randomly assigned to two parallel groups of treatment, receiving an oral daily dose of cetirizine 20 mg/die or placebo, in the morning for 3 days.

Patients underwent CPT at the LPR provoking dose (PD), before and after treatment. The last medication dose was administered in the morning, 3 h prior to CPT. PD was defined as the allergen dilution able to elicit a clinical LPR.

Subjects

P. judaica-sensitive subjects, suffering from seasonal allergic rhinoconjunctivitis were screened in the off-pollen season. The patients had a history of pollen allergy for at least two previous seasons, positive skin prick test and RAST for the specific pollen. They had no other ocular diseases and did not wear contact lenses. Patients with previously documented allergy to the studied drug, women of childbearing potential, or lactating were excluded. No topical or systemic drugs were allowed for 1 month before and during the study period. All subjects gave their informed consents and the trial was approved by the Department Ethical Committee.

Methodology

P. judaica extract (kindly provided by Bayropharm-DHS, Milan, Italy) used for skin prick test and stored as a lyophilised powder at 4 °C, was reconstituted in the diluent, saline human albumin (0.03%) (kindly provided by Bayropharm-DHS, Milan, Italy), within 24 h before test. The concentration was expressed as AUR/ml (AUR = Activity Units by RAST, Standardised Bayropharm-DHS unit) and seven dilutions were employed for the 'screening' CPT: 20, 60, 200, 600, 1,200, 2,000 and 5,000 AUR/ml. The same batch of allergen was used throughout the study. Saline albumin diluent was administered into the contralateral eye to exclude possible nonspecific reactivity. Ten milliliters of diluted allergen was instilled into the lower conjunctival sac in the right eye. The initial allergen concentration was the one identified by the skin end point test. During the following weeks, the next more concentrated allergen dilutions were used until a late clinical reaction was elicited (total symptom score more than 7 was considered positive for LPR) [24]. In the left eye 10 ml of diluent was instilled as control.

Clinical conjunctival evaluation and conjunctival scrapings were perfomed at baseline, 30 min, 6 and 24 h after challenge. The clinical reaction to the allergen was evaluated by conjunctival hyperemia, lacrimation, itching-burning and swelling of eyelids; the score was assigned according to an arbitrary four-point rating scale, from 0 to 3 (0 = absent, 1 = mild, 2 = moderate, 3 = severe). A total symptom score was calculated as the sum of individual symptom scores, as previously described [24, 29–31, 42]. Concomitantly, a conjunctival scraping for cytological and immunocytochemical assessment was obtained from both eyes immediately before, 30 min, 6 and 24 h after CPT. After topical anesthesia (ossibuprocaine 4 mg/ml, one drop each eye), the upper tarsal conjunctiva was scraped by a sterile Kimura spatula. Specimens were spread on glass slides, air-dried, stained with May-Grünwald-Giemsa dye and read by microscope (Leitz Laborlux D microscope, 500 × focus). The slides were examined separately by two investigators masked to the identity of the samples. The number of inflammatory cells, i.e. neutrophils, eosinophils, lymphocytes, was counted separately and as a total number of cells in each microscopic field; the data are expressed as a mean of 10 fields [24, 29–31, 42]. Immunoenzymatic alkaline phosphatase-monoclonal anti-alkaline phospatase (APAAP) complex procedure modified from Cordell and co-workers was employed [24, 38, 42]. Specimens were air-dried at room temperature for 30 min and incubated with appropriate dilution of purified CD54 mAb (1 mg/ml, 84H10, IgG_1 – Immunotech, Marseille) [43] or anti-cytokeratin mAb (aCK19, DAKO, Milan, Italy). After washing in phosphate-buffered saline (PBS) pH = 7.6, samples were incubated with rabbit antimouse Ig, followed by the APAAP complex. The specimens then were incubated with a substrate solution containing basic new fuchsin, napthol As biphosphate and levamisole as an inhibitor of endogenous alkaline phosphatase (Sigma, St. Louis, Mo., USA). In control samples either mAb or antimouse Ig was omitted. As negative isotype control for CD54 staining on epithelial cells, an anti-T lymphocyte (CD3) mAb OKT3, IgG_1 (Ortho Diagnostic, Raritan, N.J., USA) at 1:20 dilution of the stock solution (provided by the manufacturer) was used. The dilutions were established on the basis of previous titration experiments. All preparations were counterstained with Carazzi's hematoxylin and examined by two investigators masked to the identity of the samples. CD54 expression on epithelial cells was graded according to a rating scale from 0 to 4, where 0 = no positive cells, 1 = mild positivity on 25% of epithelial cells, 2 = mild positivity on 75% of epithelial cells, 3 = intense positivity on 75% of epithelial cells, and 4 = very intense positivity on all epithelial cells as previously described [24, 42].

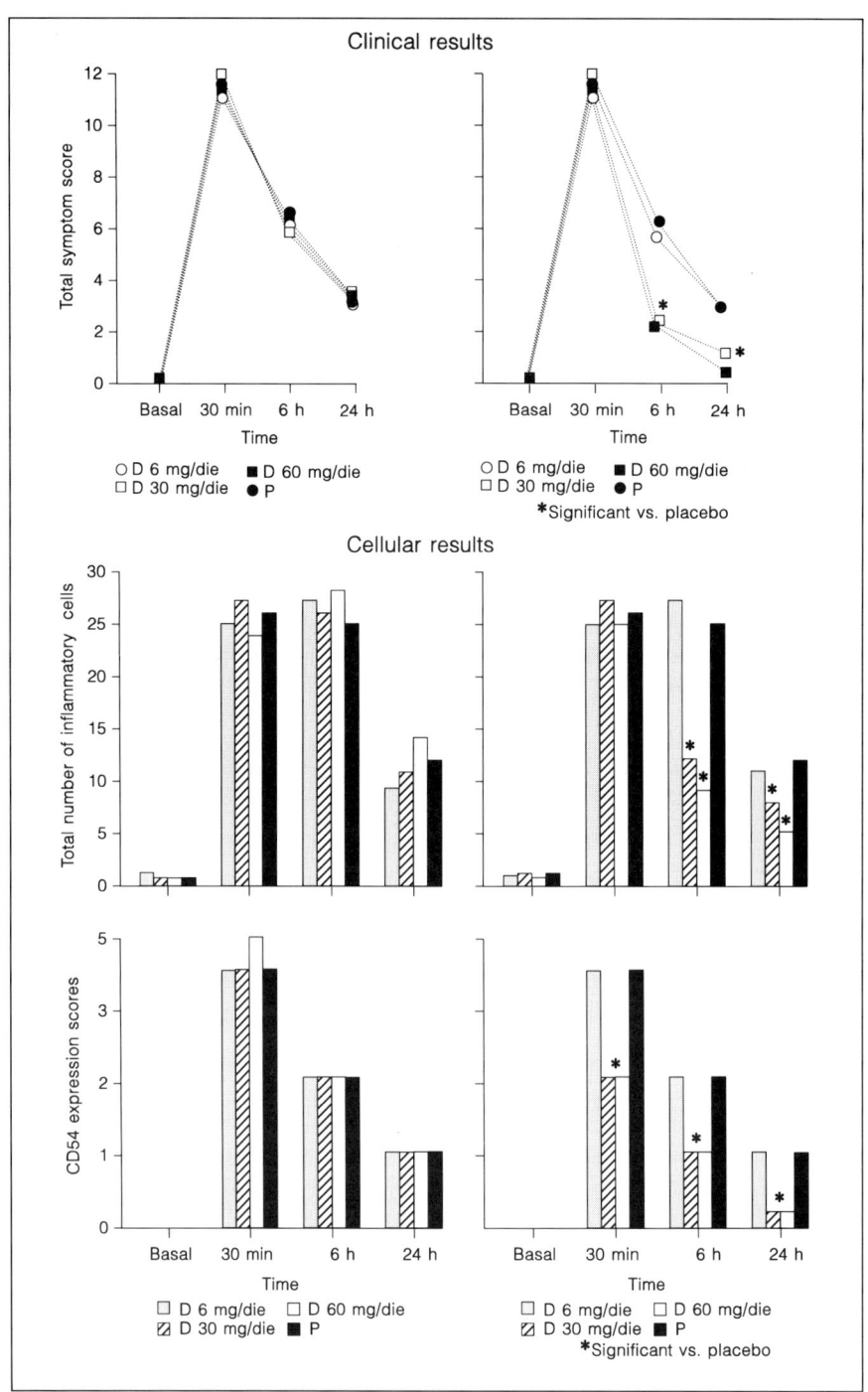

Results

As far as Deflazacort is concerned, the results may be summarised as follows (fig. 4): (a) Clinical data: the LPR was significantly reduced by D at 30 and 60 mg ($p < 0.001$). (b) Cellular data: the LPR total number of inflammatory cells was significantly reduced by D at 30 and 60 mg ($p < 0.05$). Neither the EPR clinical events nor the EPR total number of inflammatory cells were modified by any D dosage. (c) CD54 expression on conjunctival epithelium was significantly reduced by D at 30 and 60 mg in both the EPR ($p < 0,01$) and LPR ($p < 0,001$).

As far as the cetirizine trial is concerned, the results may be summarised as follows (fig. 5): (a) Clinical data: the EPR (30 min) and the LPR (6 and 24 h) were significantly reduced by C (respectively, $p < 0.004$, $p < 0.0002$ and $p<0.0003$) compared to the placebo group. (b) Cellular data: the EP (30 min) and the LPs (6 and 24 h) were significantly reduced by C (respectively, $p < 0.01$, $p < 0.002$ and $p < 0.002$) compared to the placebo group. (c) CD54 expression on conjunctival epithelium: the EP (30 min) and the LPs (6 and 24 h) were significantly reduced by C (respectively, $p < 0.008$, $p < 0.01$ and $p < 0.03$) compared to placebo group.

Conclusions

The conjunctival changes following specific CPT represent a safe and reproducible model of experimental allergic conjunctivitis. CPT prompted us to assess pharmacological modulation of the allergic response employing several drugs commonly employed in the treatment of allergic diseases, i.e. terfenadine, loratadine and more recently, Deflazacort and cetirizine [29–31, 42]. Deflazacort has an impressive protective effect on LPR events induced by CPT as well as demonstrated for several corticosteroid compounds in other models (i.e. bronchial, nasal and skin provocation test) [42]. Cetirizine exerts a significant protective effect both on EPR and LPR events induced by CPT [in preparation]. Moreover, both Deflazacort and cetirizine affected CD54 expression on epithelial cells in the EPR and LPR [42; in preparation]. This effect may suggest a new possible mechanism of action exerted by Deflazacort and cetirizine on the inflammatory process. Noteworthy is the possibility of a corticosteroid and an antihistamine to exert a relevant inhibitory effect on CD54 expression on conjunctival epithelium both on

Fig. 4. Clinical and cytological changes before (left) and after (right) CPT, and before (left) and after (right) treatment with Deflazacort. Modified from Ciprandi et al. [42].

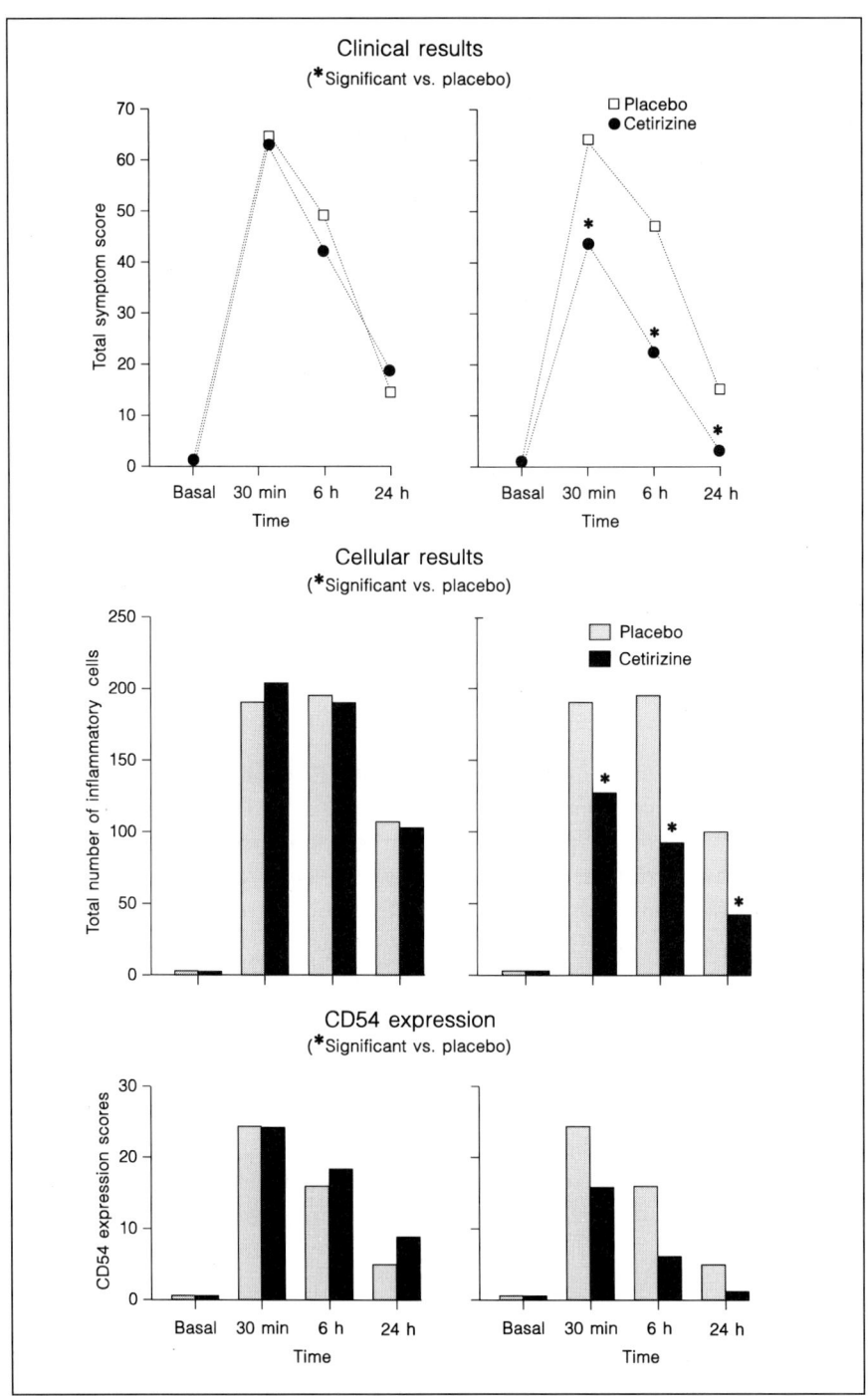

early and late-phase reactions following allergen-specific challenge. This finding is of interest and may reflect an immediate and prolonged anti-inflammatory effect of these drugs, which could in turn interfere with interactions among inflammatory cells and epithelium throughout the surface molecules. Furthermore, cetirizine has recently been shown to inhibit eosinophilic inflammation at bronchial level in allergic subjects upon allergen stimulation [44]. On the other hand, Rédier et al. [44] provided a possible experimental explanation for this observation, since they demonstrated cetirizine to be able to block intercellular adhesion between eosinophils and endothelial cells, which is the first event for eosinophilic locomotion to the allergic reaction site. Along these lines, cetirizine does not affect neutrophil adhesion to endothelial cells. We previously demonstrated that cetirizine does not exert any action on lymphocytes as far as proliferation and surface molecules expression (including ICAM-1/CD54) are concerned [34]. Altogether these data would suggest a selective action of cetirizine on some cellular populations, thus blocking the intercellular adhesion among cells types deeply involved either in allergic inflammation or in tissue damage. Interest might arise from the clinical relevance of the minimal persistent inflammation, since it might be a password for viruses to trigger the mucosa of allergic subjects. A more detailed definition of this minimal persistent inflammation, both concerning the causes and the mechanisms leading to this event, might provide a further useful tool in allergy management. A new insight for a better understanding of the molecular events of allergic inflammation might derive from the detection on epithelial cells of both ICAM-1/CD54 and LFA-1, which might play in a 'bi-directional adhesion'.

There is clearly aim for much more work on the characterization and regulation of adhesion machinery expression in the allergic response and its possible modulation by therapeutic agents which should be designed to decrease adhesion activity. For these reasons we have just started the in vivo evaluation of mizolastine, a new potent H_1 antagonist, with clinical efficacy in the treatment of allergic rhinoconjunctivitis and chronic urticaria and with antiallergic properties.

Acknowledgements

This study has been realized with the support of P.F. CNR FATMA SP2 grant to G.W. Canonica, MD, No. 5710, CNR Target project 'Ingegneria genetica' – P.F. 99 No. 91.00018, grant to M. Bagnasco by the ARMIA (Associazione Ricerca Malattie Allergiche e Immunologiche) foundation.

We thank Dr. Rina Miriello for her skillful help in preparation of the manuscript.

Fig. 5. Clinical and cytological changes before (left) and after (right) CPT, and before (left) and after (right) treatment with cetirizine. Modified from Ciprandi et al., submitted.

References

1. Montefort S, Herbert CA, Robinson C, Holgate ST: The bronchial epithelium as a target for inflammatory attack in asthma. Clin Exp Allergy 1992;L2:511.
2. Kay AB: Mediators and inflammatory cells in inflammatory diseases. Ann Allergy 1987;59:35.
3. Proud D, Sweet J, Stein P, Settipane RA, Kagey-Sobotka A, Friedlaender MH, Lichtenstein LM: Inflammatory mediator release on conjunctival provocation of allergic subjects with allergen. J Allergy Clin Immunol 1990;85:896–905.
4. Bonini SE, Bonini ST, Berruto A, Tomassini M, Carlesino S, Bucci MG, Balsano F: Conjunctival provocation test as a model for the study of allergy and inflammation in humans. Int Arch Appl Immunol 1989;88:144.
5. Kay AB: Asthma and inflammation. J Allergy Clin Immunol 1991;87:893.
6. Albelda AM: Endothelial and epithelial cell adhesion molecules. Am J Respir Cell Biol 1991;4:195.
7. Bierer BE, Burakoff SJ: T cell adhesion molecules. FASEB J 1988;2:2584.
8. Dustin ML, Springer TA: Role of lymphocyte adhesion receptors in transient interactions and cell locomotion. Annu Rev Immunol 1991;9:27.
9. Dustin ML, Rothlein R, Bhan AK, Dinarello CA, Springer TA: Induction by IL-1 and interpheron-gamma: tissue distribution, biochemistry and function of a natural adherence molecule (ICAM-1). J Immunol 1986;137:245.
10. Mantovani A, Bussolino F, Dejana E: Cytokine regulation of endothelial cell function. FASEB J 1992;6:25.
11. Ofosu-Appiah W, Warrington RJ, Morgan K, Wilkins JA: Lymphocyte extracellular matrix interactions. Induction of interferon by connective tissue components. Scand J Immunol 1989;29:1517–1521.
12. Luscinskas FW, Cybulsky MI, Kiely JM, Peckins CS, Davis VM, Gimbrone MA Jr: Cytokine-activated human endothelial monolayers support enhanced neutrophil transmigration via a mechanism involving both endothelial-leukocyte adhesion molecule-1 and intercellular adhesion molecule-1. J Immunol 1991;146:1617.
13. Kyan-Aung U, Haskard DO, Poston RN, Thornhill MH, Lee TH: Endothelial leukocyte adhesion molecule-1 and intercellular adhesion molecule-1 mediated the adhesion of eosinophils to endothelial cells in vitro and are expressed by endothelium in allergic cutaneous inflammation in vivo. J Immunol 1991;146:521.
14. Leung DYM, Poder JS, Cotran RS: Expression of Endothelial-Leukocyte Adhesion Molecule-1 in elicited late phase allergic reactions. J Clin Invest 1991;87:1805–1809.
15. Campbell AM, Vignola AM, Chanez P, Michel FB, Bousquet J, Godard PH: Bronchial epithelial cell activation in asthmatic subjects: expression of surface markers and mediator release; in Godard P, Bousquet J, Michel FB (eds): Advances in Allergology and Clinical Immunology. New York, Parthenon, 1992; pp 215–224.
16. Editorial: Adhesion molecules in diagnosis and treatment of inflammatory diseases. Lancet 1990;336:1351–1352.
17. Wegner CD, Gundel RH, Reilly P, Haynes N, Letts LG, Rothlein R: Intercellular adhesion molecule-1 (ICAM-1) in the pathogenesis of asthma. Science 1990;247:456–460.
18. Gundel RH, Wegner CD, Torcellini CA, Letts LG: The role of intercellular adhesion molecule-1 in chronic airway inflammation. Clin Exp Allergy 1992;22:569.
19. Hansel TT, Walker C: The migration of eosinophils into the sputum of asthmatics: The role of adhesion molecules. Clin Exp Allergy 1992;22:345.
20. Ciprandi G, Buscaglia S, Tosca MA, Canonica GW: Evidence of minimal persistent inflammation at conjunctival level in rhinitic patients with allergy to *Dermatophagoides pteronyssinus*. 8th Int Congr Immunology, Budapest, August 1992, vol 47, p 279.
21. Chapman MD, Heymann PW, Wilkins SR, Brown MJ, Platts-Mills TAE: Monoclonal immunoassays for major dust mite *(Dermatophagoides)* allergens, Der p I and Der f I, and quantitative analysis of the allergen content of mite and house dust extracts. J Allergy Clin Immunol 1987;80:184–194.
22. Cockcroft DW: Therapy for airway inflammation in asthma. J Allergy Clin Immunol 1991;87:914.

23 Barnes PJ: Future drug therapy for asthma. Clin Exp Allergy 1991;21:80.
24 Ciprandi G, Buscaglia S, Pesce GP, Villaggio B, Bagnasco M, Canonica GW: Allergic subjects express Intercellular Adhesion Molecule 1 (ICAM-1 or CD54) on epithelial cells of conjunctiva after allergen challenge. J Allergy Clin Immunol 1993; in press.
25 Greve JM, Davis G, Meyer AM, Forte CP, Yost SC, Marlor CW, Kamarck ME, McClelland A: The major human rhinovirus receptor is ICAM-1. Cell 1989;56:839.
26 Staunton DE, Merluzzi VJ, Rothlein R, Barton R, Marlin SD, Springer TA: A cell adhesion molecule, ICAM-1, is the major surface receptor for rhinoviruses. Cell 1989;56:849.
27 Staunton DE, Dustin ML, Erikson HP, Springer TA: The LFA 1 and rhinovirus binding sites of ICAM-1 and arrangement of its Ig-like domains. Cell 1990;61:243.
28 Pattermore PK, Johnston SL, Bardin PG: Viruses as precipitants of asthma symptoms. I. Epidemiology. Clin Exp Allergy 1992;22:325.
29 Ciprandi G, Buscaglia S, Iudice A, Canonica GW: Protective effect of terfenadine at different dosage on conjunctival provocation test. Allergy 1992;47:309–312.
30 Ciprandi G, Buscaglia S, Pesce GP, Marchesi E, Canonica GW: Protective effect of Loratadine on conjunctival provocation test. Int Arch Clin Immunol 1991;96:344–347.
31 Ciprandi G, Buscaglia S, Marchesi E, Danzig M, Kuss F, Canonica GW: Protective effect of loratadine on late phase reaction induced by conjunctival provocation test. Int Arch Allergy Appl Immunol 1992; in press.
32 Avioli LV, Gennari C, Imbimbo B: Effetti dei glucocorticoidi e loro conseguenze biologiche: Progressi in terapia glucocorticoidea. Amsterdam, Elsevier/Excerpta Medica, 1986.
33 Corticosteroids: Their Biologic Mechanism and Application to the Treatment of Asthma. An International Conference, Vienna 1988. London Royal Society of Medicine Services Limited, 1989.
34 Canonica GW, Parodi MN, Boero F, Jing G, Pesce GP, Bagnasco M: Effects of cetirizine and deflazacort on cell adhesion molecules expression on lymphocytes and epithelial cells. Schweiz Med Wochenschr 1991;40:38.
35 Campoli-Richards DM, Buckley MMT, Fitton A: Cetirizine: A review of its pharmacological properties and clinical potential in allergic rhinitis, pollen-induced asthma, and chronic urticaria. Drugs 1990;40:762–781.
36 Ciprandi G, Buscaglia S, Cerqueti PM, Canonica GW: Drug treatment of allergic conjunctivitis: A review of the evidence. Drugs 1992;43:154.
37 Friedlaender MH: Conjunctival provocation test: A model of human ocular allergy. Trans Am Ophthalmol Soc 1990;87:577–597.
38 Cordell JL, Falini B, Erber WN, Ghosh AK, Abdulaziz Z, McDonald S, Pulford KAF, Stein H, Mason DY: Immunoenzymatic labelling of monoclonal antibodies using immune complexes of alkaline phosphatase and monoclonal anti-alkaline phosphatase (APAAAP complexes). J Histochem Cytochem 1984;32:219–245.
39 Durham SR: The significance of late responses in asthma. Clin Exp Allergy 1991;21:3–7.
40 Editorial: Cutaneous responses. A window on inflammatory processes. Clin Exp Allergy 1992;22:3–6.
41 Naclerio RM, Proud D, Togias AG, Adkinson NF, Meyers DA, Kagey-Sobotka A, Plaut M, Norman PS, Lichtenstein LM: Inflammatory mediators in late antigen-induced rhinitis. N Engl J Med 1985;313:65–70.
42 Ciprandi G, Buscaglia S, Pesce GP, Iudice A, Bagnasco M, Canonica GW: Deflazacort protects late phase but not early phase events induced by allergen specific conjunctival provocation test. Allergy 1993; in press.
43 Olive D, Charmot D, Dubreuil P: Human lymphocytes functional antigens; in Feldman M (ed): Human T-Cell Clones. A New Approach to Immunoregulation. Clifton, Humana Press, 1986, p 173.
44 Rediér H, Chanez P, De Vos C, Rifai N, Clauzel AM, Michel FB, Godard P: Inhibitory effect of cetirizine on the bronchial eosinophil recruitment induced by allergen inhalation challenge in allergic patients with asthma. J Allergy Clin Immunol 1992;90:215–224.

G.W. Canonica, MD, Allergy and Clinical Immunology Service, V.le Benedetto XV, 6,
I–16132 Genoa (Italy)

Current Issues in the Therapy of Allergy and Asthma

Panel Discussion

Moderator: *S.Z. Langer* (Paris, France)
Panelists: *M.K. Church* (Southampton, UK)
G.C. Folco (Milan, Italy)
J.P. Rihoux (Braine-l'Alleud, Belgium)

Throughout the meeting we have been discussing several issues that we are going to address during the panel discussion. One is the profile of activity of histamine antagonists, and the anti-asthmatic potential of anti-allergic drugs; the question of sedation and other side effects, particularly cardiovascular and more specifically the QT interval effects, in relation to the therapeutic doses will also be discussed.

Rihoux: The main question concerns the activity in asthma of each antihistaminic. We tested ceterizine in seasonal asthma first, and we could show a clinical effectiveness of ceterizine at a dose of 20 mg/day. We showed especially that the patients who took ceterizine could decrease their daily intake of β_2-agonists and we also showed that respiratory function was better with ceterizine than with placebo.

Church: I wonder whether, in fact, any antihistaminic would today be the first-line choice in the treatment of asthma?

Rihoux: I think it is surely not the first-line choice, and it is not for treating severe asthma, just in patients with mild asthma, and especially young patients.

Do we have to treat all the patients with the same medication? Sometimes asthma is more peripheral in the bronchi, and sometimes it is more central. I think for instance that exercise-induced asthma is more central than peripheral, and it was a theory proposed perhaps 10–15 years ago by MacFadden in the United States, concerning the localization of the bronchospasm in asthmatic patients; he suggested at that time that it could be advantageous to treat patients with peripheral airway obstruction with H_1 histamine antagonists, and when the bronchospasm is more central, with atropine or anticholinergic drugs. So we have to choose a population of patients which could benefit from a special pharmacological approach, and not use the same approach in all the patients.

Church: Asthma is not one disease. In most cases, it is really two events taking place at the same time: one is the presence of an atopic status; which can manifest itself in the skin, the nose, the intestine, the eyes, or the lungs. Not all people who have hay fever have asthma and not all people that have dermatitis have asthma. Therefore, asthma must also be a lung problem in that the lung appears to be very susceptible to the allergic state. Probably, these two aspects are controlled quite differently at the genetic level. Also, the expression of the two aspects may be quantitatively different. As a consequence, there are people who are driven mainly by atopy who have basically healthy lungs and therefore have a very allergic lung disease – that is one form of asthma. At the other end of the spectrum there are people who basically have no demonstrable atopic condition at all, but have very weak lungs. They constitute non-atopic asthmatics. Among those there are many situations where it is possible to provoke an asthmatic attack of different strength. So, there are people with T-cell driven, people with mast-cell driven, and people with possibly neuronally driven asthma, each of whom may need a different type of therapy. If you have all of these manifestations then of course one would need a whole spectrum of drug therapy and it is very clear that steroids are extremely good for atopic conditions and for atopic asthma, but there are many patients whose asthma is controlled very poorly with steroids. So, even the drug that we say is by far the best doesn't always work. Likewise, β-stimulants, which are effective in the majority of patients, are again not effective in the whole population. Cromoglycate is also very effective in a few patients but not to any great degree in the majority. So, none of the drugs we have at the moment is completely satisfactory and any move to market a new drug, particularly targeted to a group of people whose treatment is not good at present, would be a great advance.

Langer: Perhaps before continuing to focus on the treatment of asthma, it would be useful to take one step back and look at the pharmacological class of H_1 antagonists as anti-allergics. Two issues were brought into the picture during this workshop: one is the separation between therapeutic effects and side effects and the second is the question of sedation. There is no evidence so far of receptor heterogeneity in the H_1 subtype, therefore the strategy has been and is clearly successful, to prevent blood-brain barrier penetration of these compounds in order to reduce or minimize sedation. It was stated clearly that there is a dose-relationship to these phenomena and that the so-called non-sedative antihistaminics can be sedative if the dose is increased sufficiently. The second question that covers side effects is related to specificity. It was clearly stated by Church that many of the early antihistaminics had an affinity for α_1-receptors, 5-HT_2 receptors, and other subtypes of receptors, among them muscarinic cholinergic, and that one way of reducing the side effects is to make drugs that are more specific, like, for instance, mizolastine. The third point at the level of conduction in the QT interval is to avoid the cardiac side effects that have been reported for some of these drugs, and I am referring now to the drugs and their active metabolites, whenever this is a valid point, for instance drugs like terfenadine and astemizole.

Rihoux: I agree completely that sedation is proportional to the dose given and it is like that for all the H_1 blockers. Curiously, a few patients are resistant to sedation, you can increase the dose in those patients; in contrast, a few patients are very sensitive and even if you give low doses you get sedation. If I have to choose, I choose a drug with a low sedating profile, but above all one which does not induce an increase in the QT interval. An important question which was raised during the meeting is that probably the non-H_1-dependent activities of the H_1 antihistamines could be of importance in treating allergy, but if you want to get these activities you have to increase the dose and so it is probably important to increase the dose without really having severe undesirable effects.

Church: The salient point is that we have, I think, in terfenadine, one of the purest H_1 blockers, or at least I thought so till the cardiac problems came along. For the treatment of allergic diseases perhaps we should take a step backwards from the specific drugs and look at some of the more 'dirty' ones, to see if some of the unwanted effects are, in fact, beneficial when treating diseases other than those involved in H_1 blockade.

Are these newer nonsedative antihistamines active when given by aerosol? Are they suitable for aerosol administration?

Rihoux: We are busy studying ceterizine for that.

Langer: Perhaps we can now focus on the question of the anti-asthmatic action, having covered, at least partly, side effects and anti-allergic efficacy. Maybe we should deal with one question, whether there is a rationale in developing H_1 antagonists for the treatment of asthma, or perhaps if this strategy has simply been, with compounds that possess the H_1 blocking effect, to explore their efficacy in asthma. Was it, strategically speaking, a rational approach based on mechanistic hypotheses, or was it an attempt to extend the indications for a compound that was otherwise established as an anti-allergic?

Histamine remains an important mediator in allergy and the blood levels of histamine are elevated in asthma patients. Moreover, there is a correlation between the plasma histamine levels and the bronchial hyperreactivity; so there are many indices showing that it is not unjustified to give an H_1 blocker in such a condition. But of course, everyone knows that histamine is only one mediator, so it is probably difficult to imagine that an antagonist for just one mediator, when you have so many mediators, will be really effective or sufficient, and it is probably the reason why the other activities, or the unexpected activities are welcome.

There is a rationale on its own for an H_1 antagonist because of the important role of histamine; however, the search for additional properties is important, like for instance the inhibition of release of other mediators of allergic reactions and the inhibition of eosinophil migration observed with nedocromil, mizolastine and cetirizine.

On the potential anti-asthmatic properties of the antihistamine drugs, one could ask whether there would be a rationale for administering an anti-H_1 drug in combination with an antileukotriene drug, and whether anyone has already tried this.

The answer is, theoretically, yes, certainly in animal screening this is the way leukotriene antagonists are examined, and I believe that during some of the relatively early trials combinations were given.

The usual acceptance of the term 'anti-asthmatic' by a clinician involves inhibition of the release of the bronchospasmic mediators. So in this respect the definition of an antihistamine cannot be accepted. If it is true that some antihistamines express an effect on the inflammatory response consequent to the allergic reaction, what is the cost-benefit in using as a supportive drug this kind of drug in the usual strategy of treating asthma? I think that the cost-benefit, if we don't consider for example the cardiotoxicity of some of these compounds, is a good one, because the side effects are almost negligible. So, from a clinical viewpoint, there is a rationale for using antihistamines but I would like to say just 'anti-allergic drugs'. We are defining the pharmacological class on the basis of selective predominant affinities and the potency at which it interacts with one receptor subtype under in vitro conditions, and this is one pharmacological criterion, but in fact the profile of activity goes far beyond this.

The ideal drug one would be looking for would be one that is prophylactic and that would prevent the onset of the chronic inflammation in the lungs. At the moment there are many possible directions, I certainly, for one, would go towards something that would interfere with cell accumulation in the airways. I think if I was having to do research in this area, I would go for something that interacted with adhesion proteins. The other target that I think is of great interest with the advent of molecular biology, is the possibility, in atopic patients at least, of specific immunotherapy. Again, this is something for the future. The third one would of course be cytokine antagonists, but that is a difficult problem, trying to find a nonpeptidic chemical antagonist for a molecule of about 10–20 kD.

Rihoux: I still consider that the profile of the corticosteroid is a good example. We have to ponder over this profile and see if it is possible to improve something in the profile. The one problem of steroids is the possibility of adaptation to them. There is some evidence now that some of the patients that are dying of 'status asthmaticus' may in fact be steroid-resistant, partly due to the therapy. Another point is the problem of hypertrophy of the smooth muscles and the collagen, and it seems that the growth factors involved in these phenomena could be really very important. But obviously, and several of the speakers made this point, it is unacceptable to develop a new class that offers no advantages over the existing therapeutic group that we have for these diseases, being allergy in one chapter and asthma in the other one.

The only way to test the validity of certain hypotheses and to test the importance of certain targets like adhesion proteins is for our medicinal chemists to design agents that are specific and selective. And, once this objective is attained, to develop a compound to reach the real test which is the human situation in pathology. And this is the only way to confirm the validity of these novel approaches, which are definitely exciting.

Subject Index

Accolate, peptido-leukotriene antagonist 80, 81
Acetaldehyde
  dehydrogenase
    activity in alcoholics 33, 41
    inhibition 33
  guinea pig effects
    blood pressure 34, 35, 39
    bradykinin 40
    histamine levels 35, 36, 39
    intratracheal pressure 34–36, 39
    mechanisms of action 39–41
    peptidase inhibitors 36, 37, 41, 42
    substance P 36–42
  pharmacological activity 40
Adhesion molecules
  allergy inflammation role 116–125
  CD54
    expression
      antihistamine effects 123–125
      house dust mite allergy 116, 117
      pollen allergy 117, 118
      rhinovirus receptor 118
Alcohol
  bronchomotor effects 33
  disulfiram treatment of alcoholism 33
  metabolism 33
Antibodies, *see also* Immunoglobin E
  production in atopic dermatitis 98
  SER 282 antibody effects
    calcium role 24, 25
    smooth muscle contraction 22–26
  therapy utilization 20
Antihistamines
  activities 1, 55, 56
  asthma therapy 109, 128
  discovery 55
  dose-dependent sedation 129
  effects
    acetaldehyde-mediated effects 37, 38, 41
    eosinophils 111, 112
    lipophilicity 110
    mast cells 110–112
    net positive charge 110
  future aims 130, 131
  investigation aspects 57, 58
  mechanisms of action 108–110
  nonsedating 45, 46, 109, 129
  potency 111
  side effects 109–111, 129, 130
  specificity 108
  structure 108
Antioxidants, chronic urticaria aggravation 52
Asthma
  antihistamine efficacy 109, 128
  clinical classification 11, 128, 129

Asthma (continued)
  pathogenesis role
    CD4 lymphocyte activation 14
    cytokines 12–15
    immunoglobin E 16
    mucosal inflammation 10–16
  specific immunotherapy 64–69
  viral induction 118
Atopic dermatitis
  allergens 97, 98
  antibody production 98
  cytokine profile 101
  definition 96, 97
  delayed cell-mediated immunity 98–100
  diagnostic criteria 103
  immunological events 100, 101
  itching role in inflammation 96, 97
  prevalence 96
  prevention measures 104
  provocative factors 101, 102
  signs 103
  skin alterations 101, 102
  treatment 105
Azelastine, *see* Antihistamines

Bradykinin
  acetaldehyde-mediated effects 40
  activities 40
Bronchoalveolar lavage fluid
  cell infiltration 13–15, 27–29
  collection 28
Burimamide, discovery 1
BWA4C, 5-lipoxygenase inhibitor 78

CD54, *see* Adhesion molecules
Ceterizine, *see also* Antihistamines
  adhesion molecule expression effects 123–125
  mechanism of action 119, 123, 125
Cyclosporin A, asthma therapy 16

Deflazacort
  adhesion molecule expression effects 121, 122
  mechanism of action 119, 123, 125
Disulfiram, alcoholism therapy 33

Eosinophils
  allergy inflammation role 107, 108
  bronchoalveolar lavage fluid infiltration 28–30
  chronic urticaria role 49–51
  effect
    antihistamines 50, 111, 112
    cytokines 30, 31
    glucocorticoids 28, 29
    protein synthesis inhibitors 30

Fibromyalgia
  symptoms 91
  treatment efficacy
    amitryptyline 92–94
    fibromyalgia treatment 91–94
    placebo 92–94
    SER 282 antibody 91–94

Glucocorticoids, therapy
  asthma 16, 119–125
  urticaria 53
Granulocyte-macrophage colony-stimulating factor
  asthma role 12, 15
  mRNA elevation in asthma 15
Guinea pig
  acetaldehyde effects 34–42
  asthma model 27, 28

Heparin, urticaria therapy 53
Histamine, *see also* Histamine receptors
  acetaldehyde effects on levels 35, 36, 39
  activity 1, 130
  asthma role 109, 130
  discovery 1
  potency 108
  thromboxane-A_2 effects 39, 40
Histamine receptors
  H_1
    agonists, *see* Antihistamines
    cloning 2
    localization 2, 3
    mediated responses 4, 5
    purification 2
    second messengers 2, 4, 5
    sequence 3
    structure 2

H_2
   discovery 1, 5
   distribution 5
   homology between species 5
   regulation 5
   second messengers 5, 6
   structure 5
H_3
   agonists 6, 7
   discovery 1, 6
   stimulation effect on airways 7, 8

ICI 207,968, 5-lipoxygenase inhibitor 77, 78
ICI D2138, 5-lipoxygenase inhibitor 78
Immunoglobin E
   asthma role 16
   inflammatory mechanisms 16
   interleukin-4 effects 16
Immunotherapy, see Specific immunotherapy
Inflammation
   diagnosis 11
   initiating factors 115
   mediators 16, 96, 97, 107, 108, 115, 116
   mucosal evidence in asthma 10, 11
   role of T lymphocytes 11–16
Interleukin-3
   asthma role 12
   mRNA elevation in asthma 15
Interleukin-4
   allergy inflammation role 107
   immunoglobin E synthesis promotion 12
Interleukin-5
   activities 12
   allergy inflammation role 107
   detection 14, 15
   mRNA elevation in asthma 15

Ketotifen, see Antihistamines

Leukotrienes, see also Lipoxygenases, Peptido-leukotrienes
   activities 71, 86
   allergy inflammation role 107

leukotriene-B_4
   activities 74, 75
   eosinophil effects 75
   receptors 74
   levels in asthmatics 73
   synthesis 71–73
Lipoxygenases
   classification 72
   5-lipoxygenase
      activating protein 72, 73, 79, 80
      calcium dependence 72
      inhibitors
         classification
            hydroxamates 78
            nonredox inhibitors 79
            redox inhibitors 77, 78
            translocation inhibitors 79, 80
         difficulty in development 76
         evaluation 77, 80, 81
         potencies 77–80
         toxicity 78
   pathways 71–73
   purification 72
   substrates 72
Loratidine, see Antihistamines

Mast cells, antihistamine effects 110–112
MK886, 5-lipoxygenase inhibitor 79, 80

Nifedipine
   calcium channel blocker 53
   urticaria therapy 53

Oxatomide, see Antihistamines

Peptido-leukotrienes
   activities 73, 74, 86
   antagonists
      binding affinities 88, 89
      development 75, 76, 87
      efficacy against asthma 80, 81
      evaluation 80, 87
   potency 73
   receptors
      characterization 73, 86
      classification 87, 88
      heterogeneity 88, 89

Platelet-activating factor, allergy inflammation role 107
Prostaglandin D_2, allergy inflammation role 107

Receptors, *see* Histamine receptors, Peptido-leukotrienes
Rhinovirus, CD54 binding 118

SER 292, *see* Antibodies
Specific immunotherapy
  asthma therapy
    efficacy 68
    house dust mite 64, 65
    indications 66–68
    mold 66
    objectives 64
    pollen 64
    safety 65, 66
  future aims 69
  history 60, 69
  mechanism of action 60, 61
  pollen-rhino-conjunctivitis therapy
    administration routes 63
    duration 63
    efficacy 62
    extract availability 63
    indications 66
    objectives 62
    safety 63
  venom immunotherapy
    duration 62
    efficacy 61
    indications 66
    safety 61
Substance P
  acetaldehyde-mediated effects 36–42
  activities 40

T lymphocytes
  CD4
    asthma role 13–15
    classification 13
    cytokine synthesis 12, 13
    effects on activity
      cyclosporin A 16
      glucocorticoids 16

Terfenadine, *see* Antihistamines
Thromboxane-A_2, histamine effects 39, 40

Ultraviolet irradiation, urticaria therapy 53
Urticaria defined
  chronic
    diagnosis 51, 52
    diet regulation 51–53
    major basic protein expression in lesions 50, 51
    occurrence 49
    pathogenesis 49–51
    treatment 52, 53
  occurrence 45
  pathogenesis mediators 44, 45
  physical types
    adrenergic urticaria 47
    aquagenic urticaria 47
    cholinergic urticaria
      lesions 47
      treatment 47
    cold urticaria
      diagnosis 46, 47
      drowning role 48
      treatment 48
    delayed dermographism 46
    localized heat urticaria 47
    pressure urticaria
      diagnosis 46
      treatment 46
    solar urticaria
      diagnosis 49
      treatment 49
    urticaria factitia
      cause 45
      diagnosis 45
      treatment 45

Venom immunotherapy, *see* Specific immunotherapy

Warfarin, urticaria therapy 53

Zileuton, 5-lipoxygenase inhibitor 78, 80, 81